The Hidden Variables Theory
The Initial Values of the Fundamental Constants in Physics

The Hidden Variables Theory

The Initial Values of the Fundamental Constants in Physics

Israel Fried

Library of Congress Cataloging-in-Publication Data:

Israel Fried, 1951-

The Hidden Variables Theory: The initial values of the Fundamental Constants in Physics

ISBN 978-1-5342-9859-0

To my dear wife Judith and my daughter Noa,
for their love and encouragement,
and
to my daughter Inbal,
for her invaluable assistance with the scientific editing
and for agreeing to take the long way to make this publication a reality.

Preface

The initial fundamental constants are the source values of the particles in the quantum level, an internal quality, but they cannot be detected directly in measurements probably due to the relative motion and mutual interaction existing between the particles in reality. In fact laboratory measurements yield the relativistic reflection of the particles, without exposing the more basic level concealed underneath. At the moment the theoretical physics research is based upon the symmetry mathematical aspect of quantum fields expressed by a combined theories associated with the gauge group $\mathbf{SU(3)} \times \mathbf{SU(2)} \times \mathbf{U(1)}$ that is considered as the Standard model of the elementary particle physics. The theory presented here uses a different approach to yield the same results and presents new discoveries. The book offers the reader the entire process of finding the theoretical fundamental constants, stretching from the quantum world to cosmology. This theory presents new topics that include new discoveries on the inner structure of the proton and the neutron with their quarks, and the source of the weak force bosons related to this inner structure, and also presents the characteristics of the Higgs boson set at 125Gev energy that its discovery was announced on 4th July 2012 at CERN.

This revision presents among other discoveries that the electric charge of any subatomic particle is originated from its Orbital Angular Momentum (OAM). If the particle's OAM is offset, the particle is neutral. No OAM no charge. A quark's OAM within baryons and mesons is a third or two thirds of the reduced Planck constant. The Cooper pair which consists of two electrons is attracting each other if one of them revolves in a direction with orbital angular momentum $(+\hbar\uparrow)$ that induces e^+ charge, and the other revolves in an opposite direction with orbital angular momentum $(-\hbar\downarrow)$ that induces e^- charge. This attraction occurs despite of the fact that the electrons possess identical intrinsic e^- charges originated from their internal structure presented in this theory which supposedly they are meant to repulse each other. In general: Two subatomic particles in contact with an opposite OAM (revolving in opposite directions) attract each other, whereas with identical OAM (revolving in the same direction) repulse each other. The proton and the electron in the hydrogen atom are acting as one entity and therefore the total OAM within the proton $(+\hbar)$ from the quarks contribution and the OAM of the electron $(-\hbar)$ at the ground state of the hydrogen atom are offset and it reflects outside a zero OAM which leads to a conclusion that the electron doesn't revolve at this orbital or any other orbitals. The Pauli Exclusion Principle stems mainly from the need of an electron at a given orbital shell with a defined OAM to join to another electron with opposite OAM that create attraction between them (OAM induces electric charge that is unassociated with the electron's intrinsic charge).

The proton, the neutron and all the baryons consist of three energy levels on which the quarks are orbiting. The third energy level is equivalent to 80.7 Gev; it plays a major role at decaying process through the weak force while it hosts charged mesons for split seconds (The Pentaquark is created this way for instance) which are emitted out through the W boson that acquires the level's energy.

The Gluons carries the force acting between the orbiting quarks at the three energy levels and decay into matter within a baryon or out of it as it is demonstrated by a jet of particles. The electron is a bound state composition of a negative Pi meson and an electron neutrino. Electrons in extremely confined space are going through a process named "spin-charge separation". The electron splits in two, a spin carrier particle named 'Spinon' and a charge carrier particle named 'Holon'.

Here the electron neutrino ν_e acting as a 'Spinon' that carries the spin and the π^- meson is acting as the 'Holon' that carries the e^- charge (originated from its intrinsic OAM presented in this theory). Research published on July 2009 by the University of Cambridge and the University of Birmingham in England showed that electrons could separate into a Spinon and a Holon.
The electron mass is a magnitude that expresses quantitatively the square of its magnetic flux quantum $\Phi_\circ^2$ (magnetic flow rate that occurs within originated from its intrinsic OAM) multiplied by a dimensionless quantity.
The neutrino ν is an Ultra relativistic state of a neutron traveling at nearly the speed of light $\mathbf{c}$
The anti neutrino $\overline{\nu}$ is an Ultra relativistic state of an anti neutron travels nearly the speed of light $\mathbf{c}$
The muon and tau neutrinos are particles that acquire their properties from travelling at different speeds near the limit of the speed of light in vacuum $\mathbf{c}$!
This theory presents a new interpretation for the second and third quarks generations which are a variations of the *down* quark symbolized d and the *up* quark symbolized u and their *anti* quarks compositions. For instance the *strange* quark s is $(u\overline{u}d)$ and the *anti strange* quark $\overline{s}$ is $(d\overline{d}\overline{d})$ or the *charm* quark c is $(d\overline{d}u)$ and the *anti charm* quark $\overline{c}$ is $(d\overline{d}\overline{u})$. This topic is widely demonstrated at the weak force decay options of the subatomic particles.
The gluons exchanged between the quarks within the baryons or being emitted out of the nucleus in the weak force process, are fused into a pair of neutrino/antineutrino particles that consists of (udd) and $(\overline{u}\overline{d}\overline{d})$ quarks in a form of a three Pi mesons π°, π^+, π^- or $\pi^\circ, \pi^\circ, \pi^\circ$ or in form of a second generation quarks (according to the new interpretation for the second generation quarks presented in this theory) as a φ meson that consists of a *strange* and an *anti strange* quarks $s\overline{s}$: $(u\overline{u}d)$ and $(d\overline{d}\overline{d})$ or a **J/psi** meson that consists of a *charm* and an *anti charm* quarks $c\overline{c}$: $(d\overline{d}u)$ and $(d\overline{d}\overline{u})$. This both mesons are often generated in the particles collision at accelerators.
The proton's charged radius is accurately resolved in this theory and it shows that it is wrongly assumed that it is the entire proton's radius. A discrepancy was found in the charge radius of the proton measured at different institutes' laboratories. The discrepancy occurs between the last recommend value $8.768\times10^{-16}[\mathrm{m}]$ of the International Council for Science's Committee on Data for Science & Technology (CODATA) set on in 2006 and a more recent result (2010) corresponding with the experimental findings of Randolf Pohl from Max Planck Institute of Quantum Optics and his colleagues, that set the charge proton radius on $8.4184\times10^{-16}[\mathrm{m}]$ in a muonic Hydrogen experiment. This theory provides a theoretical explanation for this discrepancy and introduces a finite theoretical result $r_{pcr} = 8.4124\times10^{-16}[\mathrm{m}]$ that corresponds with the findings of Randolf Pohl and his colleagues.
This theory also discovered that there must be two kinds of speed of light. The known speed of light in vacuum which this theory yield $c = 2.99792344\times10^8[\mathrm{m\,s^{-1}}]$ compared to the experiment which is $c = 2.99792458\times10^8[\mathrm{m\,s^{-1}}]$ that travels over long distances, and a speed which is a bit higher than it $c = 2.99993069\times10^8[\mathrm{m\,s^{-1}}]$ within the atomic domain setting this theory as a **non local theory**!
This revision also includes a section in chapter 12 on the Quantum numbers obtained from the Schrödinger equation that can be obtained through a different approach, not through solving the wave function! This method provides a broader understanding of them.
The higher speed of light within the atomic domain has significance in the cosmology singularity issue. The topics mentioned here and more are presented in this comprehensive book.

Contents

PART 1
The Key Constant β

1
The Proton Spinning Velocity and the Key Constant β

Ernst Rutherford's experiment of bombing the atom nucleus with alpha particles paved the road to the understanding of the atom structure and of the nuclear dimensions.[4] It is known that the volume of a nucleus is proportional to its mass number A, the number of nucleons it contains. By assuming the spherical shape of the nucleus with radius R, the following relation is formulated:

(1) $$R = R_{\circ}\mathrm{A}^{1/3}$$

$R_{\circ}$ is the estimated experimentally received value
$R_{\circ} \approx 1.2\times10^{-15}[m]$

Placing this value in Eq. 1, along with the hydrogen atom (A=1) gives its proton radius as

(2) $$r_p = R_{\circ}\mathrm{A}^{1/3} = R_{\circ} \approx 1.2\times10^{-15}\ [m]$$

The *angular momentum* of a hydrogen atom (proton) can be expressed as

$$m_p v_p r_p = \frac{h}{2\pi}$$

or

(3) $$v_p = \frac{h}{2\pi\ r_p\ m_p}$$

In order to get an estimated magnitude for the hydrogen proton spinning velocity v_p, substitute the following experimentally received values into Eq. 3:

Planck constant[2] experimentally received value: $h = 6.626176\times10^{-34}[J\ s]$
Proton mass[2] experimentally received value: $m_p = 1.6726485\times10^{-27}\ [Kg]$
Proton radius experimentally received value: $r_p \cong 1.2\times10^{-15}[m]$
Yields:

$$v_p = 5.254085881 \times 10^7 [m\, s^{-1}]$$

This proton spinning velocity magnitude is almost one fifth of the experimentally received value[2] of the speed of light in vacuum c:

$$c = 2.99792458\times10^8 [m\ s^{-1}]$$

This leads to the assumption that the spinning velocity is great enough to create a relativistic effect. The *relativistic proton radius* $r_p^{'}$ can now be calculated according to the Lorentz-Fitzgerald contraction equation:[4]

$$L = L_{\circ}\sqrt{1 - v^2 / c^2}$$

Thus, the Lorentz-Fitzgerald contraction of the proton radius takes the following form:

(4) $\quad r_p^{'} = r_p\sqrt{1 - v_p^2/c^2}$

Note: r_p is non relativistic value and it is actually considered as $r_{\circ}$ the rest radius!

So Eq. 4 implies that the hydrogen proton radius in Eq. 3 should be $r_p\sqrt{1 - v_p^2/c^2}$ rather than r_p !

Substituting Eq. 4 into Eq. 3, and raising it to a power of 2, yields:

(5) $$v_p^2\left(1 - \frac{v_p^2}{c^2}\right) = \frac{h^2}{4\pi^2\, r_p^2\, m_p^2}$$

Note: m_p is a relativistic value, $m_p = \dfrac{m_{\circ}}{\sqrt{1 - v_p^2/c^2}}$, and $m_{\circ}$ is considered as the rest mass!

I have presented the expressions $r_{\circ}$ and $m_{\circ}$ here only for clarifications, and not for further use!

Eq. 5 is a simple quadratic equation solved for v_p:

$$v_p^4 - v_p^2 c^2 + \frac{h^2\, c^2}{4\pi^2\, r_p^2\, m_p^2} = 0$$

Applying $v_p^2 = x$, gives

$$x^2 - xc^2 + \frac{h^2\, c^2}{4\pi^2\, r_p^2\, m_p^2} = 0 \Rightarrow x_{1,2} = \frac{c^2 - \sqrt{c^4 - \dfrac{h^2\, c^2}{\pi^2\, r_p^2\, m_p^2}}}{2}$$

Substituting the experimentally received values into the equation, gives the following estimated magnitude for v_p:

$v_p \approx 5.3394559 \times 10^7 \; [m/s]$

At this point, note the known constant in physics, the *fine structure constant*[3,7] α, which can be expressed as

(6) $\quad \alpha = \dfrac{v_e}{c}$

v_e is the electron velocity at the *Bohr radius*, and c is the speed of light in vacuum.

Similarly, a constant based on the proton spinning velocity, the *key constant* β, can be defined as

(7) $\quad \beta = \dfrac{v_p}{c}$

In order to produce an estimated magnitude for the *key constant* β, substitute the estimated magnitude of the proton spinning velocity, $v_p \approx 5.3394559 \times 10^7 [m/s]$, and the experimentally received value of the speed of light in vacuum *c*, into Eq. 7 to yield

$$\beta \approx 0.1781$$

The *Key constant* β as defined, presents a significant function in the mathematical development of the hidden variables theory, as will be shown in the next chapters. The *fine structure constant* α is known also as the electromagnetic coupling constant. Similarly, it seems that the **strong coupling constant** α_s was set in nuclear physics as 2/3 of the *key constant* β:

$$\alpha_s = 2/3\beta \approx 2/3 \times 0.1781 \approx 0.118733$$

This value of the **strong coupling constant** $\alpha_s \approx 0.118733$ agrees with the experiments results conducted in laboretries!

A more accurate value can be obtained once we get the exact value of the key constant from this theory. The fraction 2/3 is not accidental; it appears very often in relation to the key constant at the equations developed in this theory. In my opinion the real *strong constant* α_s should be β itself!

The **strong coupling constant** α_s is used to calculate the interactions within the atomic nucleus and between sub atomic particles.

* See a more accurate approach in the theory developed later pages 129-138, that validates these results!

2
The Elementary Charge e

The definition of the key constant β has led to the assumption of the possibility of an interaction between this magnitude and other known magnitudes in the atom domain. In order to further develop this notion, select the elementary charge e as the starting point.

It is well known that the elementary charge e is the base for the force acting on the electron in its interaction with the nucleus. Theoretically, this force derives from the exchange of virtual photons between the electron and the nucleus, and the kinetic and potential energies that are related to the electron at the Bohr radius are essentially the exchanged photons' energy. In the hydrogen atom, the electron revolves around the nucleus at the Bohr radius, allowing the exchange of a photon between them to occur with great speed and at all possible angles. This movement *seemingly* creates a continuous spherical shell around the nucleus at the Bohr radius, which resembles the effect of receiving a continuous image on a television screen resulting from the pixels lighting, which occurs rapidly. This description depicts the photon as an object moving within a spherical volume, similar to the kinetic theory of gases, also discussed in detail by Beiser.[4]

What if it is possible to describe the virtual exchanged photons' movement in terms of ideal gas molecules in a given volume?

The average kinetic energy for each gas molecule is

$$PV = \frac{1}{3} E_k$$

where V is the volume within the spherical shell, and P is the pressure that the virtual photons impose on the spherical shell. The electron kinetic energy at the Bohr radius is

$$(8) \qquad E_k = \frac{1}{2} m_e v_e^2$$

m_e is the electron mass, and v_e is the electron velocity.

v_e, can also be described using the *fine structure constant* α from Eq. 6:

$$\alpha = \frac{v_e}{c}$$

Substituting Eq. 6 and Eq. 8 into the kinetic energy expression gives

$$(9) \qquad PV = \frac{1}{3} \left[\frac{1}{2} m_e (c\alpha)^2 \right]$$

Pressure is defined as the force acting on a given surface. If the spherical shell surface is $A = 4\pi a_e^2$, and a_e is the Bohr radius, then the pressure is

(10) $$P = \frac{F}{4\pi a_e^2}$$

F is the total force acting by the virtual photons' imposed on the inner surface of the spherical shell. Substituting Eq. 10 into Eq. 9 gives

(11) $$FV = \frac{1}{3}\left(\frac{1}{2}m_e(c\alpha)^2\right)4\pi a_e^2$$

The same mathematical development can be applied for the potential energy,[4] defined as

$$PV = \frac{1}{3}\left(\frac{1}{2}E_p\right)$$

At the Bohr radius, the kinetic energy magnitude is equal to *half* of the potential energy magnitude. The potential energy E_p of the electron at the Bohr radius can also be described, based on Coulomb law,[4] as

(12) $$E_p = \left(\frac{e^2}{4\pi\,\varepsilon_\circ\,a_e}\right)$$

$\varepsilon_\circ$ is the *permittivity of vacuum constant*, e is the *elementary charge*, and a_e is the *Bohr radius*. Substituting Eq. 10 and Eq. 12 into the potential energy expression gives

(13) $$FV = \frac{1}{3}\left(\frac{e^2}{8\pi\,\varepsilon_\circ\,a_e}\right)4\pi\,a_e^2 \rightarrow FV = e^2\left(\frac{a_e}{6\varepsilon_\circ}\right)$$

Equating Eq. 11 and Eq. 13 gives

(14) $$e^2\left(\frac{a_e}{6\varepsilon_\circ}\right) = \frac{m_e\,v_e^2\,4\pi\,a_e^2}{6}$$

The right side of Eq. 14 expresses the interaction between the spherical shell and the photons' energy, which it contains. In my opinion, the equation which calculates the magnitude of the elementary charge e, should include natural magnitudes only, such as mass, velocity, and distance. The permittivity of vacuum $\varepsilon_\circ$ was historically defined by Coulomb law as the element that connects the natural magnitudes to the acting force between the charges (proton-electron). I have therefore assumed that $\varepsilon_\circ$ is derived from the terms that already exist in the equations, as received from the ideal gas model, and as described previously.

I now arrive at the crucial point in my theory development. Looking at Eq. 14, an idea came into my mind, allowing me to conduct a small thinking experiment. I assumed that the relation which multiplies the elementary charge (e^2) component in the left side of the equation, fulfills

(15) $$\left(\frac{a_e}{6\varepsilon_\circ}\right) = 1$$

or

$$a_e = 6\varepsilon_\circ$$

According to known values received from experiments, the ratio in Eq. 15 equals 0.996, very close to **1**. Looking at the difference, 0.04, the following explanation occurred to me: Perhaps there is a basic equation, which is identical in form and components to Eq. 14, but with the magnitudes of these components as their initial values, which differ from their known experimental magnitudes, and if these initial values are substituted into Eq. 14, a result which fulfills the assumption in Eq. 15 is received.

This hypothesis is the heart of the entire theory presented here, and can be extended to all fundamental constants simply by realizing that they all have *basic equations* containing their initial values. I have learned, through further development of this concept that these initial values do exist, and they lay *hidden* within the known experimental values. Applying the assumption from Eq. 15 to Eq. 14 yields the following expression for the elementary charge e

(16) $$e^2_{elementary} = \frac{m_e v_e^2 4\pi a_e a_e}{6} = m_e v_e^2 4\pi a_e \varepsilon_\circ$$

And based on Eq. 5

$$v_p^2\left(1-\frac{v_p^2}{c^2}\right) = \frac{h^2}{4\pi^2 r_p^2 m_p^2}$$

it is possible to present the *Planck constant* h in two forms: for the electron,

(17a) $$h = m_e v_e 2\pi a_e$$

and for the proton,

(17b) $$h = m_p v_p 2\pi r_p \sqrt{1-\beta^2}$$

The expression for the proton charge is identical to the expression for the electron charge from Eq. 16, and can be obtained by equating Eq. 17a with Eq. 17b, while raising both equations to a power of **2** and dividing them by $(6\pi m_e)$:

(18) $$e^2_{elementary} = \frac{m_p v_p^2 4\pi r_p^2 (1-\beta^2)}{6}\left(\frac{m_p}{m_e}\right)$$

Multiplying the right side of Eq. 18 by $\frac{a_e}{a_e}$, and applying the relation $v_p = c\beta$ from Eq. 7, gives

(19) $$e^2_{elementary} = \left(\frac{m_p c^2 \beta^2 4\pi r_p a_e \sqrt{1-\beta^2}}{6}\right)\left(\frac{m_p r_p \sqrt{1-\beta^2}}{m_e a_e}\right)$$

I have deliberately separated the components on the right side of Eq. 19.

Now, equating Eq. 17a to Eq. 17b, applying the relations $v_e = c\alpha$ from Eq. 6 and $v_p = c\beta$ from Eq. 7, along with reducing the $(2\pi c)$ component, yields

(20) $$m_e \alpha a_e = m_p \beta r_p \sqrt{1-\beta^2}$$

Transposing the terms in Eq. 20, yields:

(21) $\alpha/\beta = m_p r_p \sqrt{1-\beta^2} / m_e a_e$

Substituting Eq. 21 into Eq. 19, with the relation $a_e = 6\varepsilon_0$ from Eq. 15, yields

(22) $e^2 = m_p v_p^2 4\pi\varepsilon_0 r_p \sqrt{1-\beta^2} \times \alpha/\beta$ or $e^2 = m_p c^2 \alpha\beta \, 4\pi\varepsilon_0 r_p \sqrt{1-\beta^2}$

- In later development I discovered a different approach that shows the relation between the ground state radius a_0 in the Hydrogen atom and the permittivity of the vacuum constant ε_0. It is also recommended to see pages 129 -138, for how to obtain the β constant differently and also find a relation between the electron mass m_e and the square of the magnetic flux quantum Φ_0^2. Here is a sample of it:

The magnetic flux quantum Φ_0 is defined according to

(1) $e = h/2\Phi_0 \rightarrow e^2 = h^2/4\Phi_0^2$

where e is the elementary charge of an electron and h is Planck's constant. The electrostatic force acting on the electron at the Bohr level is

(2) $e^2/4\pi a_0^2 \varepsilon_0 = m_e v_e^2 / a_0 \rightarrow e^2 = m_e v_e^2 4\pi a_0 \varepsilon_0$

where a_0 is the Bohr radius, ε_0 is the vacuum permittivity, m_e is the electron mass, and v_e is the electron velocity at the Bohr radius. The electron's angular momentum at the Bohr radius is

(3) $h/2\pi = m_e v_e a_0 \rightarrow h = 2\pi m_e v_e a_0$

By substituting Eq. (3) in Eq. (1) (the squared term), we obtain

(4) $e^2 = m_e v_e 2\pi a_0 \times m_e v_e 2\pi a_0 / 4\Phi_0^2$

We can then rewrite Eq. (4) as follows

(5) $e^2 = m_e v_e^2 4\pi a_0 \times (\pi a_0 m_e / 4\Phi_0^2)$

By multiplying both the numerator and denominator of Eq. (5) by ε_0, we have

(6) $e^2 = m_e v_e^2 4\pi a_0 \varepsilon_0 (\pi a_0 m_e / 4\varepsilon_0 \Phi_0^2)$

According to Eq. (2), the expression in parentheses in Eq. (6) should equal unity; thus, we obtain

(7) $(\pi a_0 m_e / 4\varepsilon_0 \Phi_0^2) = 1$

$$\frac{\pi}{4} \times \frac{5.291\,772\,109 \times 10^{-11}\ \text{m}}{8.854\,187\,812\,8 \times 10^{-12}\ \text{A}^2\text{s}^4\text{kg}^{-1}\text{m}^{-3}} \times \frac{9.109\,383\,701\,5 \times 10^{-31}\ \text{kg}}{4.275\,936\,823 \times 10^{-30}\ \text{kg}^2\text{m}^4\text{s}^{-4}\text{A}^{-2}} = 1$$

The parameters values are from NIST CODATA 2018! Now, consider the multiplication of the vacuum permittivity and the square of the magnetic flux in the denominator of Eq. (7), and multiply their units

(8) $(\text{A}^2\text{s}^4\text{kg}^{-1}\text{m}^{-3})(\text{kg}^2\text{m}^4\text{s}^{-4}\text{A}^{-2}) = \text{kg}\,\text{m}$

Analysis of Equations (7) and (8).

a. After reducing the units in Eq. (8), we obtained units corresponding to those in the numerator in Eq. (7). The result in Eq. (8) implies two options: Either Φ_0^2 has units of mass $[\text{kg}]$ or units of length $[\text{m}]$ or, vice versa, ε_0 has units of mass $[\text{kg}]$ or has units of length $[\text{m}]$.

b. The orders of magnitude in Eq. (7) indicate the scale of magnitude of the Bohr radius scale is 10^{-11}m, and the vacuum permittivity scale is 10^{-12}Fm. The scale of the electron mass is 10^{-31}kg, and the scale of the square of the magnetic flux quantum is $10^{-30}\,\text{Wb}^2$. From this consideration and the two options presented in (a), it is nonsensical for ε_0 to have units of mass or Φ_0^2 to have units of length. No particles with a mass on the scale of 10^{-12} kg or a length on the scale of 10^{-30}m exist in the atom domain. We can conclude that the Bohr radius a_0 and the vacuum permittivity ε_0 are the same entity: $\varepsilon_0 \equiv a_0$, and the electron mass m_e and the square of the magnetic flux quantum Φ_0^2 are the same entity: $m_e \equiv \Phi_0^2$.

PART 2

The Hidden Initial Values of the Fundamental Constants

1
The Initial Values Concept

In the development for the elementary charge e seen in the previous chapter, I have mentioned the concept of the *hidden initial values* of the fundamental constants in relation to the basic equations that are similar in form to the known equations, using the known experimental values.

These initial values are called so due to the fact that they do not include the influence of the proton-electron interaction in the atomic domain. The interaction influence is based on the relativistic factors, originating from the mutual movement of the proton and the electron around a common mass center in the hydrogen atom. Each constant is based on a hidden initial value, and by entering the interaction influence, the constant receives its *measured experimental value*. In order to distinguish the initial value of a constant from its known experimental value, each initial value will be marked with the tilde sign (~) above it. For example, the initial value of the electron mass will be marked as $\tilde{m}_e$.

At this point, it is necessary to emphasize two parameters that are exceptional to this initial value marking: the key constant β and the speed of light in vacuum c. The magnitude of the key constant β, as it became clear during development of this theory, is a value that does not change, even under the proton-electron interaction influence. The speed of light in vacuum c is described throughout the development as the magnitude *matching the experiment only,* which comes from the post-factum understanding of the development. Therefore, trust that this clarification will be given later in the development, after laying the mathematical basis for the theoretical expression calculating the speed of light in vacuum.

Note also that, since β and c are defined as exceptions for the initial value marking, they directly influence the marking of the proton spinning velocity v_p through Eq. 7:

$$\beta = \frac{v_p}{c}$$

Therefore, the spinning velocity v_p does not change during the development and is not marked as an initial value.

2
The Initial Values of the Mass Ratio $\tilde{m}_p / \tilde{m}_e$ and the Fine Structure Constant $\tilde{\alpha}$

In previous sections, the expressions for calculating the hidden initial values of the ratio between the proton mass and the electron mass $\tilde{m}_p / \tilde{m}_e$ and of the fine structure constant $\tilde{\alpha}$ were developed. Both are *dimensionless characters*. I demonstrate here two possible development courses, in order to yield the desired expressions:

Course 1:

Equating Eq. 16 and Eq. 18, while applying $\tilde{v}_e = c\tilde{\alpha}$ from Eq. 6 in its initial form, and $v_p = c\beta$ from Eq. 7, yields

(23) $$\frac{\tilde{m}_e c^2\, \tilde{\alpha}^2 4\pi\, \tilde{a}_e^2}{2\times 3} = \frac{\tilde{m}_p c^2 \beta^2 4\pi\, \tilde{r}_p^2 (1-\beta^2)}{2\times 3}\left(\frac{\tilde{m}_p}{\tilde{m}_e}\right)$$

Dividing both sides of Eq. 23 by $\left(\tilde{m}_e c^2 \tilde{a}_e \tilde{r}_p\right)$, and reducing the **2** from the numerator, gives

(24) $$\left(\frac{\tilde{a}_e}{\tilde{r}_p}\right)\left(\frac{2\pi\tilde{\alpha}^2}{3}\right) = \left(\frac{\tilde{m}_p^2}{\tilde{m}_e^2}\right)\left(\frac{\tilde{r}_p}{\tilde{a}_e}\right)\left(\frac{2\pi\beta^2\left(1-\beta^2\right)}{3}\right)$$

Multiplying both sides of Eq. 24 by its right side expression gives

(25) $$\left(\frac{\tilde{m}_p^2}{\tilde{m}_e^2}\right)\frac{4\pi^2\, \tilde{\alpha}^2\beta^2\left(1-\beta^2\right)}{9} = \left(\frac{\tilde{m}_p^4}{\tilde{m}_e^4}\right)\left(\frac{\tilde{r}_p^2}{\tilde{a}_e^2}\right)\frac{4\pi^2\beta^4\left(1-\beta^2\right)^2}{9}$$

This mathematical operation allows for the isolation of only the desired unknown variables $\tilde{m}_p / \tilde{m}_e$ and $\tilde{\alpha}$ on the left side of Eq. 25. One way of rearranging the left side of Eq. 25 is

(26) $$\left[\left(\frac{\tilde{m}_p}{\tilde{m}_e}\right)\frac{2\pi\, \tilde{\alpha}^2\sqrt{1-\beta^2}}{3}\right]\left[\left(\frac{\tilde{m}_p}{\tilde{m}_e}\right)\frac{2\pi\beta^2\sqrt{1-\beta^2}}{3}\right] = \left(\frac{\tilde{m}_p^4}{\tilde{m}_e^4}\right)\left(\frac{\tilde{r}_p^2}{\tilde{a}_e^2}\right)\frac{4\pi^2\beta^4\left(1-\beta^2\right)^2}{9}$$

Another possible layout is

(27) $$\left[\left(\frac{\tilde{m}_p}{\tilde{m}_e}\right)\frac{2\pi\tilde{\alpha}\beta\sqrt{1-\beta^2}}{3}\right]\left[\left(\frac{\tilde{m}_p}{\tilde{m}_e}\right)\frac{2\pi\tilde{\alpha}\beta\sqrt{1-\beta^2}}{3}\right] = \left(\frac{\tilde{m}_p^4}{\tilde{m}_e^4}\right)\left(\frac{\tilde{r}_p^2}{\tilde{a}_e^2}\right)\frac{4\pi^2\beta^4\left(1-\beta^2\right)^2}{9}$$

The following expressions can be derived from these two forms, enabling one to find the desired unknown variables:

a. $\left(\frac{\tilde{m}_p}{\tilde{m}_e}\right)\frac{2\pi\tilde{\alpha}^2\sqrt{1-\beta^2}}{3}$

b. $\left(\frac{\tilde{m}_p}{\tilde{m}_e}\right)\frac{2\pi\beta^2\sqrt{1-\beta^2}}{3}$

c. $\left(\frac{\tilde{m}_p}{\tilde{m}_e}\right)\frac{2\pi\tilde{\alpha}\beta\sqrt{1-\beta^2}}{3}$

The choice of these expressions can be justified by comparing each to a specific solution that fulfills the following value limitations:

1. The mass ratio value measured by experiment:[2]

$$\frac{m_p}{m_e} \approx 1836$$

2. The fine structure constant α value known from experiments:[3]

$$\alpha \approx \frac{1}{137}$$

3. The key constant β based on the estimated value:

$$\beta \approx 0.1781$$

In order to find the desired unknown variables, numerically substitute these limitations into the expressions, in order to guess the form of the solution. The best-suited solution found was
$\left(\sqrt{\beta/\tilde{\alpha}}\right)^n$,
$n = 0, \pm1, \pm2, \pm3, \ldots$

It is important to state here, that the suggested solution incorporates $\tilde{\alpha}$ and β as found in the original equations, Eq. 16 and Eq. 18, and there is no use of any added variables or unexplained arbitrary numbers!

Course 2:

Dividing both sides of Eq. 23 by $\left(\tilde{m}_e c^2\ \tilde{a}_e \tilde{r}_p \pi^2\right)$, and reducing the **2** from the denominator, gives

$$\left(\frac{\tilde{a}_e}{\tilde{r}_p}\right)\frac{4\tilde{\alpha}^2}{3\pi} = \left(\frac{\tilde{m}_p^2}{\tilde{m}_e^2}\right)\left(\frac{\tilde{r}_p}{\tilde{a}_e}\right)\frac{4\beta^2\left(1-\beta^2\right)}{3\pi} \tag{28}$$

Multiplying both sides of Eq. 28 by its right side expression yields

(29) $$\left(\frac{\tilde{m}_p^2}{\tilde{m}_e^2}\right)\frac{16\beta^2\tilde{\alpha}^2\left(1-\beta^2\right)}{9\pi^2}=\left(\frac{\tilde{m}_p^4}{\tilde{m}_e^4}\right)\left(\frac{\tilde{r}_p^2}{\tilde{a}_e^2}\right)\frac{16\beta^4\left(1-\beta^2\right)^2}{9\pi^2}$$

Again, this allows the isolation of the desired unknown variables $\tilde{m}_p/\tilde{m}_e$ and $\tilde{\alpha}$ on the left side. One way of rearranging the left term in Eq. 29 is

(30) $$\left[\left(\frac{\tilde{m}_p}{\tilde{m}_e}\right)\frac{4\tilde{\alpha}^2\sqrt{1-\beta^2}}{3\pi}\right]\left[\left(\frac{\tilde{m}_p}{\tilde{m}_e}\right)\frac{4\beta^2\sqrt{1-\beta^2}}{3\pi}\right]=\left(\frac{\tilde{m}_p^4}{\tilde{m}_e^4}\right)\left(\frac{\tilde{r}_p^2}{\tilde{a}_e^2}\right)\frac{16\beta^4\left(1-\beta^2\right)^2}{9\pi^2}$$

Another possible form is

(31) $$\left[\left(\frac{\tilde{m}_p}{\tilde{m}_e}\right)\frac{4\tilde{\alpha}\beta\sqrt{1-\beta^2}}{3\pi}\right]\left[\left(\frac{\tilde{m}_p}{\tilde{m}_e}\right)\frac{4\tilde{\alpha}\beta\sqrt{1-\beta^2}}{3\pi}\right]=\left(\frac{\tilde{m}_p^4}{\tilde{m}_e^4}\right)\left(\frac{\tilde{r}_p^2}{\tilde{a}_e^2}\right)\frac{16\beta^2\left(1-\beta^2\right)^2}{9\pi^2}$$

Again, the following expressions can be derived from these two forms, allowing us to find the desired unknown variables:

a. $\left(\frac{\tilde{m}_p}{\tilde{m}_e}\right)\frac{4\tilde{\alpha}^2\sqrt{1-\beta^2}}{3\pi}$

b. $\left(\frac{\tilde{m}_p}{\tilde{m}_e}\right)\frac{4\beta^2\sqrt{1-\beta^2}}{3\pi}$

c. $\left(\frac{\tilde{m}_p}{\tilde{m}_e}\right)\frac{4\tilde{\alpha}\beta\sqrt{1-\beta^2}}{3\pi}$

As in the development of course 1, compare each expression to a specific numerical solution, using the same limitations. The form of the solution received from substituting the values into these expressions, is identical to the form received by course 1.

At this point, it is possible to obtain more expressions using similar development courses from Eq. 23 simply by dividing the terms in different variations, and isolating the desired unknown variables. Although these developments are not shown here due to their length, I have included them in Table 1, which presents the expressions by a sequence based on the suggested solution $\left(\sqrt{\beta/\tilde{\alpha}}\right)^n$, depending on the ascending and descending of the power ***n***.

Table 1. The Solutions Sequence for $\tilde{m}_p / \tilde{m}_e$

Ordinal number	**Column a**	**Column b**	***n***
1	$\left(\frac{\tilde{m}_p}{\tilde{m}_e}\right)\frac{\left(1-\beta^2\right)}{3}=\left(\sqrt{\beta/\tilde{\alpha}}\right)^4$	$\left(\frac{\tilde{m}_p}{\tilde{m}_e}\right)\frac{4\beta^3\sqrt{1-\beta^2}}{3\pi\,\tilde{\alpha}}=\left(\sqrt{\beta/\tilde{\alpha}}\right)^4$	**4**
2	$\left(\frac{\tilde{m}_p}{\tilde{m}_e}\right)\frac{8\beta^5}{3\tilde{\alpha}}=\left(\sqrt{\beta/\tilde{\alpha}}\right)^3$	$\left(\frac{\tilde{m}_p}{\tilde{m}_e}\right)\frac{2\pi\beta^2\sqrt{1-\beta^2}}{3}=\left(\sqrt{\beta/\tilde{\alpha}}\right)^3$	**3**
3	$\left(\frac{\tilde{m}_p}{\tilde{m}_e}\right)\frac{\tilde{\alpha}\left(1-\beta^2\right)}{3\beta}=\left(\sqrt{\beta/\tilde{\alpha}}\right)^2$	$\left(\frac{\tilde{m}_p}{\tilde{m}_e}\right)\frac{4\beta^2\sqrt{1-\beta^2}}{3\pi}=\left(\sqrt{\beta/\tilde{\alpha}}\right)^2$	**2**
4	$\left(\frac{\tilde{m}_p}{\tilde{m}_e}\right)\frac{8\beta^4}{3}=\left(\sqrt{\beta/\tilde{\alpha}}\right)^1$	$\left(\frac{\tilde{m}_p}{\tilde{m}_e}\right)\frac{2\pi\,\tilde{\alpha}\beta\sqrt{1-\beta^2}}{3}=\left(\sqrt{\beta/\tilde{\alpha}}\right)^1$	**1**
5	$\left(\frac{\tilde{m}_p}{\tilde{m}_e}\right)\frac{\tilde{\alpha}^2\left(1-\beta^2\right)}{3\beta^2}=\left(\sqrt{\beta/\tilde{\alpha}}\right)^0$	$\left(\frac{\tilde{m}_p}{\tilde{m}_e}\right)\frac{4\tilde{\alpha}\beta\sqrt{1-\beta^2}}{3\pi}=\left(\sqrt{\beta/\tilde{\alpha}}\right)^0$	**0**
6	$\left(\frac{\tilde{m}_p}{\tilde{m}_e}\right)\frac{8\beta^3\tilde{\alpha}}{3}=\left(\sqrt{\beta/\tilde{\alpha}}\right)^{-1}$	$\left(\frac{\tilde{m}_p}{\tilde{m}_e}\right)\frac{2\pi\,\tilde{\alpha}^2\sqrt{1-\beta^2}}{3}=\left(\sqrt{\beta/\tilde{\alpha}}\right)^{-1}$	**-1**
7	$\left(\frac{\tilde{m}_p}{\tilde{m}_e}\right)\frac{\tilde{\alpha}^3\left(1-\beta^2\right)}{3\beta^3}=\left(\sqrt{\beta/\tilde{\alpha}}\right)^{-2}$	$\left(\frac{\tilde{m}_p}{\tilde{m}_e}\right)\frac{4\tilde{\alpha}^2\sqrt{1-\beta^2}}{3\pi}=\left(\sqrt{\beta/\tilde{\alpha}}\right)^{-2}$	**-2**
8	$\left(\frac{\tilde{m}_p}{\tilde{m}_e}\right)\frac{8\beta^2\tilde{\alpha}^2}{3}=\left(\sqrt{\beta/\tilde{\alpha}}\right)^{-3}$	$\left(\frac{\tilde{m}_p}{\tilde{m}_e}\right)\frac{2\pi\,\tilde{\alpha}^3\sqrt{1-\beta^2}}{3\beta}=\left(\sqrt{\beta/\tilde{\alpha}}\right)^{-3}$	**-3**
9	$\left(\frac{\tilde{m}_p}{\tilde{m}_e}\right)\frac{\tilde{\alpha}^4\left(1-\beta^2\right)}{3\beta^4}=\left(\sqrt{\beta/\tilde{\alpha}}\right)^{-4}$	$\left(\frac{\tilde{m}_p}{\tilde{m}_e}\right)\frac{4\tilde{\alpha}^3\sqrt{1-\beta^2}}{3\pi}=\left(\sqrt{\beta/\tilde{\alpha}}\right)^{-4}$	**-4**

Note here expression number **5** in **column b**, which will be used in further developments:

(32) $$\left(\frac{\tilde{m}_p}{\tilde{m}_e}\right)\frac{4\tilde{\alpha}\beta\sqrt{1-\beta^2}}{3\pi}=\left(\sqrt{\beta/\tilde{\alpha}}\right)^0$$

Using the same development courses, isolate the ratio between the initial Bohr radius and the initial proton radius $\tilde{a}_e/\tilde{r}_p$ from Eq. 23.

This produces the following Table 2, which also presents the expressions by a sequence based on the suggested solution $\left(\sqrt{\beta/\tilde{\alpha}}\right)^n$, depending on the ascending and descending of the power ***n***:

Table 2. The Solutions Sequence for $\tilde{a}_e/\tilde{r}_p$

Ordinal number	Column a	Column b	n
1	$\left(\frac{\tilde{a}_e}{\tilde{r}_p}\right)\frac{4\beta^2}{3\pi}=\left(\sqrt{\beta/\tilde{\alpha}}\right)^4$	$\left(\frac{\tilde{a}_e}{\tilde{r}_p}\right)\frac{\tilde{\alpha}\sqrt{1-\beta^2}}{3\beta}=\left(\sqrt{\beta/\tilde{\alpha}}\right)^4$	**4**
2	$\left(\frac{\tilde{a}_e}{\tilde{r}_p}\right)\frac{2\pi\ \tilde{\alpha}\beta}{3}=\left(\sqrt{\beta/\tilde{\alpha}}\right)^3$	$\left(\frac{\tilde{a}_e}{\tilde{r}_p}\right)\frac{8\beta^4}{3\sqrt{1-\beta^2}}=\left(\sqrt{\beta/\tilde{\alpha}}\right)^3$	**3**
3	$\left(\frac{\tilde{a}_e}{\tilde{r}_p}\right)\frac{4\tilde{\alpha}\beta}{3\pi}=\left(\sqrt{\beta/\tilde{\alpha}}\right)^2$	$\left(\frac{\tilde{a}_e}{\tilde{r}_p}\right)\frac{\tilde{\alpha}^2\sqrt{1-\beta^2}}{3\beta^2}=\left(\sqrt{\beta/\tilde{\alpha}}\right)^2$	**2**
4	$\left(\frac{\tilde{a}_e}{\tilde{r}_p}\right)\frac{2\pi\ \tilde{\alpha}^2}{3}=\left(\sqrt{\beta/\tilde{\alpha}}\right)^1$	$\left(\frac{\tilde{a}_e}{\tilde{r}_p}\right)\frac{8\ \tilde{\alpha}\beta^3}{3\sqrt{1-\beta^2}}=\left(\sqrt{\beta/\tilde{\alpha}}\right)^1$	**1**
5	$\left(\frac{\tilde{a}_e}{\tilde{r}_p}\right)\frac{4\tilde{\alpha}^2}{3\pi}=\left(\sqrt{\beta/\tilde{\alpha}}\right)^0$	$\left(\frac{\tilde{a}_e}{\tilde{r}_p}\right)\frac{\tilde{\alpha}^3\sqrt{1-\beta^2}}{3\beta^3}=\left(\sqrt{\beta/\tilde{\alpha}}\right)^0$	**0**
6	$\left(\frac{\tilde{a}_e}{\tilde{r}_p}\right)\frac{2\pi\ \tilde{\alpha}^3}{3\beta}=\left(\sqrt{\beta/\tilde{\alpha}}\right)^{-1}$	$\left(\frac{\tilde{a}_e}{\tilde{r}_p}\right)\frac{8\ \tilde{\alpha}^2\beta^2}{3\sqrt{1-\beta^2}}=\left(\sqrt{\beta/\tilde{\alpha}}\right)^{-1}$	**-1**
7	$\left(\frac{\tilde{a}_e}{\tilde{r}_p}\right)\frac{4\tilde{\alpha}^3}{3\pi\beta}=\left(\sqrt{\beta/\tilde{\alpha}}\right)^{-2}$	$\left(\frac{\tilde{a}_e}{\tilde{r}_p}\right)\frac{\tilde{\alpha}^4\sqrt{1-\beta^2}}{3\beta^4}=\left(\sqrt{\beta/\tilde{\alpha}}\right)^{-2}$	**-2**
8	$\left(\frac{\tilde{a}_e}{\tilde{r}_p}\right)\frac{2\pi\ \tilde{\alpha}^4}{3\beta^2}=\left(\sqrt{\beta/\tilde{\alpha}}\right)^{-3}$	$\left(\frac{\tilde{a}_e}{\tilde{r}_p}\right)\frac{8\ \tilde{\alpha}^3\beta}{3\sqrt{1-\beta^2}}=\left(\sqrt{\beta/\tilde{\alpha}}\right)^{-3}$	**-3**
9	$\left(\frac{\tilde{a}_e}{\tilde{r}_p}\right)\frac{4\tilde{\alpha}^4}{3\pi\beta^2}=\left(\sqrt{\beta/\tilde{\alpha}}\right)^{-4}$	$\left(\frac{\tilde{a}_e}{\tilde{r}_p}\right)\frac{\tilde{\alpha}^5\sqrt{1-\beta^2}}{3\beta^5}=\left(\sqrt{\beta/\tilde{\alpha}}\right)^{-4}$	**-4**

Note here two of the expressions in Table 2, which will be used later:

Expression number **5** in **column a**:

(33) $$\left(\frac{\tilde{a}_e}{\tilde{r}_p}\right)\frac{4\tilde{\alpha}^2}{3\pi}=\left(\sqrt{\beta/\tilde{\alpha}}\right)^0$$

and expression number **5** in **column b**:

(34) $$\left(\frac{\tilde{a}_e}{\tilde{r}_p}\right)\frac{\tilde{\alpha}^3\sqrt{1-\beta^2}}{3\beta^3}=\left(\sqrt{\beta/\tilde{\alpha}}\right)^0$$

It is possible to test the accuracy of Eq. 33 and Eq. 34 by equating Eq. 33 to Eq. 32:

(35) $$\left(\frac{\tilde{a}_e}{\tilde{r}_p}\right)\frac{4\tilde{\alpha}^2}{3\pi} = \left(\frac{\tilde{m}_p}{\tilde{m}_e}\right)\frac{4\tilde{\alpha}\beta\sqrt{1-\beta^2}}{3\pi}$$

This result in the *Planck constant* h, as shown in Eq. 17, with the $(2\pi c)$component reduced:

$$\tilde{m}_e\tilde{\alpha}\,\tilde{a}_e = \tilde{m}_p\beta\,\tilde{r}_p\sqrt{1-\beta^2}$$

Similarly, the equating process between Eq. 34 and Eq. 32 yields the *Planck constant* h again. Now, by dividing expression **2** from **column b** in Table 1 by expression **3** from the same column, it is possible to generate the $\beta/\tilde{\alpha}$ ratio:

$$\left(\frac{\tilde{m}_p}{\tilde{m}_e}\right)\left(\frac{2\pi\beta^2\sqrt{1-\beta^2}}{3}\right)\left(\frac{\tilde{m}_e}{\tilde{m}_p}\right)\left(\frac{3\pi}{4\beta^2\sqrt{1-\beta^2}}\right) = \sqrt{\beta/\tilde{\alpha}}$$

After reducing, it yields

$$\frac{\pi^2}{2} = \sqrt{\beta/\tilde{\alpha}}$$

or after raising both sides of the equation to a power of **2,** yields

(36) $$\frac{\pi^4}{4} = \frac{\beta}{\tilde{\alpha}}$$

In a similar way, dividing expression **1** from **column a** in **Table 1** by expression **2** from **column b** of the same table, yields

$$\left(\frac{\tilde{m}_p}{\tilde{m}_e}\right)\left[\frac{(1-\beta^2)}{3}\right]\left(\frac{\tilde{m}_e}{\tilde{m}_p}\right)\left(\frac{3}{2\pi\beta^2\sqrt{1-\beta^2}}\right) = \sqrt{\beta/\tilde{\alpha}}$$

$$\frac{\sqrt{1-\beta^2}}{2\pi\beta^2} = \sqrt{\beta/\tilde{\alpha}}$$

or after raising both sides of the equation to a power of **2,** yields

(37) $$\frac{(1-\beta^2)}{4\pi^2\beta^4} = \frac{\beta}{\tilde{\alpha}}$$

Comparing the result from Eq. 36 to the result of Eq. 37 yields

$$\sqrt{1-\beta^2} = \pi^3\beta^2$$

or

(38) $$\beta = \left[\frac{\left(-1+\sqrt{1+4\pi^6}\right)}{2\pi^6}\right]^{1/2}$$

Eq. 38 can deduce the theoretical value of the *key constant* β using only the value of π !

Substituting $\pi = 3.1415926535$ in Eq. 38 yields

$$\beta = 0.178145016$$

This allows the calculation of the γ factor (the Lorentz-Fitzgerald contraction[4]), which relates to the proton spinning velocity:

$$\gamma = \sqrt{1-\beta^2} = 0.9840042445$$

It also allows the calculation of the initial value of the *fine structure constant* $\tilde{\alpha}$ using Eq. 36, as here:

$$\tilde{\alpha} = \frac{4\beta}{\pi^4}$$

to receive

$$\tilde{\alpha} = 7.315334297 \times 10^{-3}$$

Knowing the initial value of the *fine structure constant* allows the calculation of the value of the $\beta / \tilde{\alpha}$ ratio:

$$\beta / \tilde{\alpha} = 24.35227277$$

Substituting all the values received so far into Eq. 32 gives the initial value of the ratio between the proton mass and the electron mass $\frac{\tilde{m}_p}{\tilde{m}_e}$:

$$\frac{\tilde{m}_p}{\tilde{m}_e} = 1837.410991$$

This mass ratio relates to the *hidden initial mass values*, and, as can be seen, it is very close to the experimentaly received value, which served as one of the limitations:

$$\frac{m_p}{m_e} \approx 1836$$

3
The Initial Value of the Gravitation Constant $\widetilde{G}$

This stage of finding the initial value for the *gravitation constant* $\widetilde{G}$ is the basis for the theoretical development of the expressions calculating the speed of light in vacuum c and the initial value of the elementary charge $\widetilde{e}$ using Planck mass and Planck length.

Starting with Eq. 22 already introduced in the previous chapter,

$$e^2_{elementary} = m_p c^2 \beta \, \alpha \; 4\pi \; \varepsilon_\circ r_p \sqrt{1-\beta^2}$$

this equation receives its initial values form as follows:

(39) $$\widetilde{e}^2_{elementary} = \widetilde{m}_p c^2 \beta \, \widetilde{\alpha} \; 4\pi \; \widetilde{\varepsilon}_\circ \widetilde{r}_p \sqrt{1-\beta^2}$$

This form is identical to the original, only containing the marked initial values. Dividing both sides of Eq. 39 by $\left(\widetilde{e}^2\right)$, and rearranging the right side, yields

(40) $$\left(\frac{1}{\widetilde{e}}\right)\left(\frac{1}{\widetilde{e}}\widetilde{m}_p \beta\right) c^2 \, \widetilde{\alpha} \; 4 \; \pi \; \widetilde{\varepsilon}_\circ \; \widetilde{r}_p \sqrt{1-\beta^2} = 1$$

The first expression in the parenthesis on the left side of Eq. 40 is the inverse value of the elementary charge $\widetilde{e}$, specifying here the *quantity of particles* (electrons, protons) in one Coulomb. This is therefore a *dimensionless expression*, which serves as a *counting tool of particles*:

$$\frac{1 \; \text{~~coulomb~~}}{\widetilde{e} \; \text{~~coulomb~~}} = \frac{1}{\widetilde{e}}$$

The second expression on the left side of Eq. 40, also in parenthesis, is defined as the hidden initial value of *Planck mass*:

(41) $$\left(\frac{1}{\widetilde{e}}\widetilde{m}_p \beta\right) = \widetilde{m}_{pl}$$

In addition, multiplying both terms of Eq. 41 by the speed of light in vacuum c gives an expression that describes the *total of momentum found in one Coulomb of protons*:

(42) $$\left(\frac{1}{\widetilde{e}}\widetilde{m}_p v_p\right) = \widetilde{m}_{pl} c$$

Substituting Eq. 41 into Eq. 40 yields

(43) $$\left(\frac{1}{\tilde{e}}\right)\tilde{m}_{pl}\, c^2\, \tilde{\alpha}\; 4\pi\, \tilde{\varepsilon}_\circ\, \tilde{r}_p \sqrt{1-\beta^2} = 1$$

Dividing both sides of Eq. 43 by $\left[\left(\frac{1}{\tilde{e}}\right)\tilde{m}_{pl}\right]^2$ and rearranging yields

(44) $$\frac{1}{4\pi\,\tilde{\varepsilon}_\circ \left(\frac{1}{\tilde{e}^2}\right)\tilde{m}_{pl}^2} = \frac{\tilde{\alpha}\, c^2\, \tilde{r}_p\sqrt{1-\beta^2}}{\left(\frac{1}{\tilde{e}}\right)\tilde{m}_{pl}}$$

The left side of Eq. 44 is the expression for the hidden initial value of the gravitation constant $\tilde{G}$ which can also be written as

(45) $$\tilde{G} = \frac{\tilde{e}^2}{4\pi\,\tilde{\varepsilon}_\circ\,\tilde{m}_{pl}^2}$$

Therefore, the right side must also equal to $\tilde{G}$:

(46) $$\tilde{G} = \frac{\tilde{\alpha}\, c^2\, \tilde{r}_p\sqrt{1-\beta^2}}{\left(\frac{1}{\tilde{e}}\right)\tilde{m}_{pl}} = \frac{\tilde{\alpha}\,\tilde{r}_p\sqrt{1-\beta^2}}{\left(\frac{1}{\tilde{e}}\right)} \times \left(\frac{c^2}{\tilde{m}_{pl}}\right)$$

The form of the right side of Eq. 46 resembles the expression for the condition of forming a black hole, which is the Schwarzschild solution to certain conditions in the general relativity theory:[5]

$$G = \frac{Rc^2}{2M} \quad or \Rightarrow \quad \frac{2GM}{Rc^2} = 1$$

Where: R is the radius of the black hole, M the black hole mass and c the speed of light in vacuum. A theoretical discussion on Schwarzschild radius from the point of view of the hidden initial values theory is presented later. From this analogy, Eq. 46 receives the following form:

(47) $$\tilde{G} = \frac{\tilde{l}_{pl}\, c^2}{\tilde{m}_{pl}}$$

The **2** which appears in the numerator of the Schwarzschild solution does not conflict with formulation of Eq. 47, since it can be included in one of the other components, such as the mass or radius, and therefore is not mentioned in the further development.

The expression that multiplies the quadratic speed of light in vacuum c in Eq. 47 is the initial value for Planck length, and can be deduced from the analogy to Eq. 46 as follows:

(48) $$\tilde{l}_{pl} = \frac{\tilde{\alpha}\,\tilde{r}_p\sqrt{1-\beta^2}}{\left(\frac{1}{\tilde{e}}\right)}$$

In short, Eq. 46, along with the definitions of Planck mass from Eq. 41 and *Planck length* from Eq. 48, connects the gravitation constant in the macro level and the quantum elements in the micro level!

As previously explained, the component $(1/\tilde{e})$ serves as a counting tool for counting length units in Eq. 48. In this case, it counts the *Coulomb quantity of Planck lengths that are contained within the proton radius*!

In conclusion, Eq. 46 includes four parameters: Planck length $\tilde{l}_{pl}$, Planck mass $\tilde{m}_{pl}$, the gravitation constant $\tilde{G}$, and the speed of light in vacuum c.

Isolation of $(\tilde{m}_{pl}^2)$ in Eq. 45, yields

$$(49) \qquad \tilde{m}_{pl}^2 = \frac{\tilde{e}^2}{4\pi\, \tilde{\varepsilon}_\circ \tilde{G}}$$

Isolation of $(\tilde{m}_{pl}^2)$in Eq. 46, yields

$$(50) \qquad \tilde{m}_{pl}^2 = \frac{\tilde{\alpha}^2 c^4 \tilde{r}_p^2 \left(1-\beta^2\right)}{\left(\frac{1}{\tilde{e}}\right)^2 \tilde{G}^2}$$

Equating Eq. 49 and Eq. 50 yields

$$(51) \qquad \frac{\tilde{e}^2}{4\pi\, \tilde{\varepsilon}_\circ \tilde{G}} = \frac{\tilde{\alpha}^2 c^4 \tilde{r}_p^2 \left(1-\beta^2\right)}{\left(\frac{1}{\tilde{e}}\right)^2 \tilde{G}^2}$$

Transposing the terms and reducing the $(\tilde{e}^2)$ and $\tilde{G}$ components in Eq. 51 results in

$$(52) \qquad \tilde{G} = 4\pi \tilde{\varepsilon}_\circ \tilde{\alpha}^2 c^4 \tilde{r}_p^2 \left(1-\beta^2\right)$$

Eq. 52 allows us to find the initial gravitation constant $\tilde{G}$ without using the $\tilde{e}$ and $\tilde{m}_{pl}$ components! It is important to notice that Eq. 52 describes the dependence between the physical parameters based on their *magnitudes only* (not by their units)!

The unit system becomes meaningless in this process from the conventional point of view. This presentation of the parameters will be extended to other parameters as well in the next chapters.

Substituting the following known Maxwell relation[4] with the initial values:

$$\varepsilon_\circ = \frac{1}{\mu_\circ c^2} \Rightarrow \tilde{\varepsilon}_\circ = \frac{1}{\tilde{\mu}_\circ c^2}$$

into Eq. 52, yields

$$(53) \qquad \tilde{G} = \frac{4\pi}{\tilde{\mu}_\circ c^2} \tilde{\alpha}^2 c^4 \tilde{r}_p^2 \left(1-\beta^2\right)$$

and substituting Eq. 33 in the following form

$$\tilde{r}_p = \tilde{a}_e \frac{4\tilde{\alpha}^2}{3\pi}$$

into Eq. 53, along with the relation $\tilde{a}_e = 6\tilde{\varepsilon}_\circ$ from Eq. 15, yields

$$(54) \qquad \tilde{G} = \frac{256\tilde{\alpha}^6\left(1-\beta^2\right)}{\pi\tilde{\mu}_\circ^3 c^2}$$

where $\tilde{\mu}_\circ$ is the initial value of the permeability of the vacuum constant. This form of Eq. 54 shows an expression that contains fundamental constants only!

4
The Expressions for the Initial Values of $\widetilde{e}, \widetilde{m}_e, \widetilde{\mu}_\circ, \widetilde{\varepsilon}_\circ$

The process of finding the initial values of the elementary charge $\widetilde{e}$, the electron mass $\widetilde{m}_e$, and the permeability of vacuum constant $\widetilde{\mu}_\circ$ along with the permittivity of vacuum constant $\widetilde{\varepsilon}_\circ$, includes three steps:

Step 1:

Substituting the relation $v_p^2 = c^2\beta^2$ from Eq. 7 into the initial form of Eq. 18 yields

$$\widetilde{e}^{\,2} = \frac{\widetilde{m}_p c^2\beta^2 4\pi\, \widetilde{r}_p^{\,2}(1-\beta^2)}{2\times 3}\left(\frac{\widetilde{m}_p}{\widetilde{m}_e}\right)$$

Rearranging Eq. 32 as follows

$$\frac{\widetilde{m}_p}{\widetilde{m}_e} = \frac{3\pi}{4\widetilde{\alpha}\beta\sqrt{1-\beta^2}}$$

And substituting it instead of the mass ratio expression into Eq. 18 while reducing as needed, gives

$$(55) \qquad \widetilde{e}^{\,2} = \frac{\widetilde{m}_p c^2\beta\pi^2\widetilde{r}_p^{\,2}\sqrt{1-\beta^2}}{2\widetilde{\alpha}}$$

Rearranging Eq. 32 to the following form

$$\widetilde{m}_p = \frac{3\pi\,\widetilde{m}_e}{4\widetilde{\alpha}\beta\sqrt{1-\beta^2}}$$

and substituting it into Eq. 55, while reducing as needed, yields

$$(56) \qquad \widetilde{e}^{\,2} = \frac{3\widetilde{m}_e c^2\beta\pi^3\widetilde{r}_p^{\,2}}{2\left(4\widetilde{\alpha}^2\beta\right)}$$

At this stage, the key constant β is not reduced in order to preserve all possible solutions!

Now, taking Eq. 33

$$\left(\frac{\widetilde{a}_e}{\widetilde{r}_p}\right)\frac{4\widetilde{\alpha}^2}{3\pi} = \left(\sqrt{\beta/\widetilde{\alpha}}\right)^0$$

and raising both sides to a power of **2** and applying the relation $\widetilde{a}_e = 6\widetilde{\varepsilon}_\circ$ from Eq. 15 yields

(57) $$\widetilde{r}_p^2 = 4\widetilde{\varepsilon}_\circ^2 \frac{16\widetilde{\alpha}^4}{\pi^2}$$

Substituting Eq. 57 into Eq. 56 yields

(58) $$\widetilde{e}^2 = \frac{3\left(16\widetilde{m}_e c^2 \beta\, \widetilde{\alpha}^2 \pi\, \widetilde{\varepsilon}_\circ^2\right)}{2\beta}$$

Substituting the known identity $\widetilde{\varepsilon}_\circ^2 = 1/\widetilde{\mu}_\circ^2 c^4$ from Maxwell relations in its initial form into Eq. 58 yields

(59) $$\widetilde{e}^2 = \frac{48\widetilde{m}_e \beta\, \widetilde{\alpha}^2 \pi}{2\beta\, c^2 \widetilde{\mu}_\circ^2}$$

Dividing both sides of Eq. 59 by $\left(\widetilde{e}^2\right)$and rearranging the right side of the equation yields

(60) $$\left(\frac{48\widetilde{\alpha}^2}{\widetilde{e}\, c^2 \beta}\right)\left(\frac{\widetilde{m}_e \pi \beta}{2\widetilde{e}\, \widetilde{\mu}_\circ^2}\right) = 1$$

The first expression in the parenthesis on the left side of Eq. 60 can be defined as follows:

$$\frac{\widetilde{e} c^2 \beta}{48\widetilde{\alpha}^2} = 1$$

This definition comes from the need to isolate one expression that includes only two unknown variables, $\widetilde{e}$ and c, versus an expression with all the remaining variables.
Substituting the following data values into the defined expression

Experimentally received value of elementary charge[2]: $e = 1.6021892 \times 10^{-19}[C]$
Experimentally received value of speed of light in vacuum[2]:
$c = 2.99792458 \times 10^8 \left[ms^{-1}\right]$
Initial fine structure constant found in theory: $\widetilde{\alpha} = 7.315334297 \times 10^{-3}$
Key constant: $\beta = 0.178145016$
Yields

(61) $$\frac{ec^2 \beta}{48\widetilde{\alpha}^2} = 0.9986$$

This result suggests the possible existence of an initial value for the elementary charge based on the same concept as Eq. 15. It is possible to rearrange the definition expression from Eq. 60 to yield additional two forms of Eq. 61:

$$c^2 = \frac{48\widetilde{\alpha}^2}{\widetilde{e}\beta}$$

$$\tilde{e} = \frac{48\tilde{\alpha}^2}{c^2\beta}$$

Substituting the values mentioned above into the following form of Eq. 61 for $\tilde{e}$

$$\tilde{e} = \frac{48\tilde{\alpha}^2}{c^2\beta}$$

Yields the *expected* initial value of the elementary charge $\tilde{e}$

$$\tilde{e} = 1.604332974 \times 10^{-19} [C]$$

This value serves as *estimation only* at this point, since it is based on the experimentally received value of the speed of light in vacuum c, and not on the value received by a theoretical calculation, hence the name *expected value*. This value will receive its final theoretical value once the theoretical value of c is determined.

By the definition in Eq. 60, the second expression in parenthesis on the left side of the equation also equals to **1**:

(62) $$\frac{2\tilde{e}\tilde{\mu}_\circ^2}{\tilde{m}_e\pi\beta} = 1$$

Rearranging this expression yields the following additional forms:

$$\frac{\tilde{e}}{\tilde{m}_e} = \frac{\pi\beta}{2\tilde{\mu}_\circ^2}$$

$$\tilde{e} = \frac{\tilde{m}_e\pi\beta}{2\tilde{\mu}_\circ^2}$$

$$\tilde{m}_e = \frac{2\tilde{\mu}_\circ^2\tilde{e}}{\pi\beta}$$

Also, reducing components in the overall form of Eq. 60 and rearranging yields

(63) $$\frac{\tilde{e}^2\tilde{\mu}_\circ^2 c^2}{24\tilde{m}_e\pi\,\tilde{\alpha}^2} = 1$$

These equations will be used in further development.

<u>**Step 2:**</u>

The following form of Eq. 61

$$\frac{\tilde{e}c^2\beta}{48\tilde{\alpha}^2} = 1$$

allows equating to Eq. 62 in the following form:

$$\frac{2\tilde{e}\tilde{\mu}_\circ^2}{\tilde{m}_e\pi\beta}=1$$

producing

$$(64)\qquad \frac{2\tilde{e}\tilde{\mu}_\circ^2}{\tilde{m}_e\pi\beta}=\frac{\tilde{e}c^2\beta}{48\tilde{\alpha}^2}$$

Placing the known identity $c^2=1/\tilde{\mu}_\circ\tilde{\varepsilon}_\circ$ from Maxwell relations in its initial form into Eq. 64, and rearranging the equation, yields

$$(65)\qquad \left(\frac{2\tilde{e}\tilde{\varepsilon}_\circ}{\tilde{m}_e\pi}\right)\left(\frac{48\tilde{\alpha}^2\tilde{\mu}_\circ^3}{\tilde{e}\beta^2}\right)=1$$

Setting the first expression in the parenthesis on the left side of Eq. 65 equal to z

$$(66)\qquad \left(\frac{2\tilde{e}\tilde{\varepsilon}_\circ}{\tilde{m}_e\pi}\right)=z$$

leads to the second expression in the parenthesis on the left side of Eq. 65 being equal to $1/z$:

$$(67)\qquad \left(\frac{48\tilde{\alpha}^2\tilde{\mu}_\circ^3}{\tilde{e}\beta^2}\right)=\frac{1}{z}$$

Multiplying Eq. 66 by the **inverse** of Eq. 67, while placing the known identity $\tilde{\varepsilon}_\circ=1/\tilde{\mu}_\circ c^2$ from Maxwell relations in its initial form, yields

$$(68)\qquad \frac{2\ \tilde{e}^2\beta^2}{48\ \tilde{m}_e\pi\ \tilde{\alpha}^2\tilde{\mu}_\circ^4c^2}=z^2$$

Multiplying Eq. 68 by $(\tilde{\mu}_\circ^2c^2)$, and rearranging the left side of the equation, yields

$$(69)\qquad \left(\frac{\tilde{e}^2\tilde{\mu}_\circ^2c^2}{24\ \tilde{m}_e\pi\ \tilde{\alpha}^2}\right)\left(\frac{\beta^2}{\tilde{\mu}_\circ^6c^4}\right)=z^2$$

The first expression in the parenthesis on the left side of Eq. 69 is Eq. 63, which is equal to **1**, and therefore yields

$$\left(\frac{\beta^2}{\tilde{\mu}_\circ^6c^4}\right)=z^2$$

or, after extracting the square root,

$$(70)\qquad \left(\frac{\beta}{\tilde{\mu}_\circ^3c^2}\right)=z$$

Step 3:

Using Eq. 44

$$\frac{1}{4\pi\,\tilde{\varepsilon}_{\circ}\left(\frac{1}{\tilde{e}^{2}}\right)\tilde{m}_{pl}^{2}}=\frac{\tilde{\alpha}c^{2}\tilde{r}_{p}\sqrt{1-\beta^{2}}}{\left(\frac{1}{\tilde{e}}\right)\tilde{m}_{pl}}$$

placing the relation $\tilde{m}_{pl}=\left(\frac{1}{\tilde{e}}\tilde{m}_{p}\beta\right)$ from Eq. 41 and multiplying both sides by $\left(\frac{4\pi\,\tilde{m}_{pl}}{\tilde{\alpha}\,c^{2}\tilde{r}_{p}}\right)$, while rearranging the equation, yields

(71) $$\frac{\tilde{e}^{3}}{\tilde{\varepsilon}_{\circ}c^{2}\tilde{m}_{p}\tilde{r}_{p}\tilde{\alpha}\beta}=4\pi\,\tilde{e}\sqrt{1-\beta^{2}}$$

Rearranging Eq. 32 as follows

$$\frac{3\pi\,\tilde{m}_{e}}{\tilde{m}_{p}4\tilde{\alpha}\beta\sqrt{1-\beta^{2}}}=1$$

and substituting the relation $\sqrt{1-\beta^{2}}=\beta^{2}\pi^{3}$ from Eq. 38, yields

(72) $$\tilde{m}_{p}=\frac{3\pi\,\tilde{m}_{e}}{4\tilde{\alpha}\beta^{3}\pi^{3}}$$

Substituting the relation $\tilde{a}_{e}=6\tilde{\varepsilon}_{\circ}$ from Eq. 15 in its initial form into Eq. 33 yields

(73) $$\tilde{r}_{p}=\frac{4\tilde{\alpha}^{2}6\tilde{\varepsilon}_{\circ}}{3\pi}$$

Multiplying Eq. 72 by Eq. 73 yields

(74) $$\tilde{m}_{p}\tilde{r}_{p}=\frac{6\tilde{\varepsilon}_{\circ}\tilde{m}_{e}\tilde{\alpha}}{\beta^{3}\pi^{3}}$$

Substituting Eq. 74 into the denominator of the left side of Eq. 71 yields

(75) $$\frac{\tilde{e}\left(\tilde{e}^{2}\beta^{2}\pi^{3}\right)}{6\tilde{m}_{e}\tilde{\varepsilon}_{\circ}^{2}c^{2}\tilde{\alpha}^{2}}=4\pi\,\tilde{e}\sqrt{1-\beta^{2}}$$

Now, substituting the known identity $\tilde{\varepsilon}_{\circ}=1/\tilde{\mu}_{\circ}c^{2}$ from Maxwell relations in its initial form into the expression for z, as defined in Eq. 66

$$\left(\frac{2\tilde{e}\,\tilde{\varepsilon}_{\circ}}{\tilde{m}_{e}\pi}\right)=z$$

and rearranging the equation, yields

(76) $$\frac{\tilde{e}}{\tilde{m}_e} = \frac{\pi\, z\, \tilde{\mu}_\circ c^2}{2}$$

Substituting Eq. 76 into the left side of Eq. 75, placing the identity $\tilde{\varepsilon}_\circ^2 = 1/\tilde{\mu}_\circ^2 c^4$ from Maxwell relations in its initial form, and rearranging gives

(77) $$\frac{\tilde{e}^2 c^4 \beta^2 \pi^4 \tilde{\mu}_\circ^3 z}{12\tilde{\alpha}^2} = 4\pi\, \tilde{e}\sqrt{1-\beta^2}$$

Substituting Eq. 61, raised to a power of **2,** as follows

$$\tilde{e}^2 c^4 = \frac{(48)^2 \tilde{\alpha}^4}{\beta^2}$$

into the left side of Eq. 77 and rearranging yields

(78) $$\left(192\pi^4 \tilde{\alpha}^2 z\right)\tilde{\mu}_\circ^3 = 4\pi\, \tilde{e}\sqrt{1-\beta^2}$$

Assuming at this point that the expression in the parenthesis on the left side of Eq. 78 is equal to **1**, meaning:

(79) $$192\pi^4 \tilde{\alpha}^2 z = 1$$

$$z = \frac{1}{192\pi^4 \tilde{\alpha}^2}$$

applying the initial value of the fine structure constant $\tilde{\alpha}$ calculated from this theory

$$\tilde{\alpha} = 7.315334297 \times 10^{-3}$$

into Eq. 79 yields the value of z

$z = 0.999150356$

The assumption made for Eq. 78 reduces its form into the following:

(80) $$\tilde{\mu}_\circ^3 = 4\pi\, \tilde{e}\sqrt{1-\beta^2}$$

Substituting the following data values into Eq. 80:

Expected initial value of the elementary charge: $\tilde{e} = 1.604332974 \times 10^{-19} [C]$

Key constant: $\beta = 0.178145016$

Yields the expected initial value of the permeability of vacuum constant $\tilde{\mu}_\circ$:

$\tilde{\mu}_\circ = 1.256513373 \times 10^{-6} [N / A^2]$

Again, this value serves as estimation only at this point, since it is based on the expected initial value of the elementary charge e, which is based on the experimental value of the speed of light in vacuum c.

The final value will be determined once the exact theoretical value of c is established.

Substituting the following data values into Eq. 70

$$\left(\frac{\beta}{\tilde{\mu}_\circ^3 c^2}\right) = z$$

Expected initial value of permeability of vacuum constant: $\tilde{\mu}_\circ = 1.256513373 \times 10^{-6} \left[N / A^2\right]$
Experimentally received value of speed of the light in vacuum:[2] $c = 2.99792458 \times 10^8 \left[ms^{-1}\right]$
Key constant: $\beta = 0.178145016$
Yields the value of z:

$z = 0.999150535$

The value received here is identical to the value received for z using Eq. 79. z depends only on the value of $\tilde{\alpha}$, which in turn depends only on β and π.
This leads to the conclusion that the assumption made in Eq. 79 stating that the expression $192\pi^4\tilde{\alpha}^2 z$ is equal to **1**, is correct. In addition, it appears that the term z from Eq. 79 is actually the γ factor, which relates to the interaction between the initial electron velocity at the Bohr radius $\tilde{v}_e = c\tilde{\alpha}$, and the proton spinning velocity $v_p = c\beta$:

(81) $$z = \frac{1}{192\pi^4\tilde{\alpha}^2} = \sqrt{1 - \tilde{v}_e^2 / v_p^2} = \sqrt{1 - \tilde{\alpha}^2 / \beta^2}$$

Substituting the following data values into $\sqrt{1 - \tilde{\alpha}^2 / \beta^2}$
Key constant: $\beta = 0.178145016$
Initial fine structure constant found in theory: $\tilde{\alpha} = 7.315334297 \times 10^{-3}$

Yields the value of the γ factor

$$\gamma_{proton\ /electron} = \sqrt{1 - \tilde{\alpha}^2 / \beta^2} = 0.999156521$$

For comparison, the value of z calculated by Eq. 79 was
$z = 0.999150356$.
Equating Eq. 70 to Eq. 81 yields

(82) $$c^2 = \frac{192\pi^4\tilde{\alpha}^2\beta}{\tilde{\mu}_\circ^3} = \frac{\beta}{\tilde{\mu}_\circ^3\sqrt{1 - \tilde{\alpha}^2 / \beta^2}}$$

Rearranging Eq. 82 and placing the known identity $c^2 = 1 / \tilde{\mu}_\circ\tilde{\varepsilon}_\circ$ from Maxwell relations in its initial form yields

$$\tilde{\mu}_\circ^3 = \frac{1}{c^2}\left(\frac{\beta}{\sqrt{1 - \tilde{\alpha}^2 / \beta^2}}\right) = \tilde{\mu}_\circ\tilde{\varepsilon}_\circ\left(\frac{\beta}{\sqrt{1 - \tilde{\alpha}^2 / \beta^2}}\right)$$

or

(83) $$\tilde{\varepsilon}_{\circ} = \tilde{\mu}_{\circ}^{2}\left(\frac{\sqrt{1-\tilde{\alpha}^{2}/\beta^{2}}}{\beta}\right)$$

Placing $\tilde{\mu}_{\circ}$ from Eq. 80 into Eq. 83 yields

(84) $$\tilde{\varepsilon}_{\circ} = \left(4\pi\,\tilde{e}\sqrt{1-\beta^{2}}\right)^{2/3}\left(\frac{\sqrt{1-\tilde{\alpha}^{2}/\beta^{2}}}{\beta}\right)$$

This is the expression for calculating the initial value of the permittivity of vacuum constant $\tilde{\varepsilon}_{\circ}$. The following chapters will further develop the equations in order to receive the calculated value.

5

The Theoretical Calculation of the Initial Values of $\tilde{e}, \tilde{m}_e, \tilde{\mu}_\circ, \tilde{\varepsilon}_\circ, \tilde{m}_p, \tilde{a}_e, \tilde{r}_p$

This chapter presents the findings of the hidden initial values of the different constants, using only the theoretically calculated initial fine structure constant $\tilde{\alpha}$ and the key constant β. Finding the initial values paves the road to calculating their final value, which matches their experimental value. This allows us to determine the known experimental values of the different constants based on a purely theoretical calculation. The process described in this chapter includes two steps:

Step 1:

Isolating $\tilde{r}_p$ from Eq. 34, while raising the equation to a power of **2**, gives

(85) $$\tilde{r}_p^2 = \tilde{a}_e^2 \frac{\tilde{\alpha}^6\left(1-\beta^2\right)}{9\beta^6}$$

Substituting the relation $\tilde{a}_e = 6\tilde{\varepsilon}_\circ$ from Eq. 15 in its initial form into the right side of Eq. 34 yields

$$\tilde{r}_p^2 = \frac{4\,\tilde{\varepsilon}_\circ^2\,\tilde{\alpha}^6\left(1-\beta^2\right)}{\beta^6}$$

Substituting Eq. 85 into Eq. 52 for the initial gravitation constant

$$\tilde{G} = 4\pi\,\tilde{\varepsilon}_\circ \tilde{\alpha}^2 c^4 \tilde{r}_p^2\left(1-\beta^2\right)$$

and substituting the known identity $\tilde{\varepsilon}_\circ = 1/\tilde{\mu}_\circ c^2$ from Maxwell relations in its initial form yields

(86) $$\tilde{G} = \frac{16\pi\,\tilde{\alpha}^8\left(1-\beta^2\right)^2}{\beta^6 c^2 \tilde{\mu}_\circ^3}$$

Substituting the relation $\left(1-\beta^2\right) = \pi^6\beta^4$ from Eq. 38 for one of the $\left(1-\beta^2\right)$ components on the right side of Eq. 86 yields

(87) $$\tilde{G} = \frac{16\pi^7\tilde{\alpha}^8\beta^4\left(1-\beta^2\right)}{\beta^6 c^2 \tilde{\mu}_\circ^3}$$

Rearranging Eq. 87 gives

(88) $$\tilde{G} = \left(2\pi^2\tilde{\alpha}^8\beta^4\sqrt{1-\beta^2}\right)\left(\frac{8\pi^5\sqrt{1-\beta^2}}{\beta^6 c^2 \tilde{\mu}_\circ^3}\right)$$

Step 2:

Substituting relation $\tilde{m}_{pl} = \left(\frac{1}{\tilde{e}}\tilde{m}_p\beta\right)$ from Eq. 41 into Eq. 46 here

$$\tilde{G} = \frac{\tilde{\alpha}\, c^2\, \tilde{r}_p\sqrt{1-\beta^2}}{\left(\frac{1}{\tilde{e}}\right)\tilde{m}_{pl}} = \frac{\tilde{\alpha}\, \tilde{r}_p\sqrt{1-\beta^2}}{\left(\frac{1}{\tilde{e}}\right)}\left(\frac{c^2}{\tilde{m}_{pl}}\right)$$

yields

(89) $$\tilde{G} = \frac{\tilde{\alpha}\, \tilde{e}^2 c^2 \tilde{r}_p\sqrt{1-\beta^2}}{\tilde{m}_p\beta}$$

In addition, rearranging Eq. 32 into the following form:

$$\tilde{m}_p = \frac{3\pi\, \tilde{m}_e}{4\tilde{\alpha}\beta\sqrt{1-\beta^2}}$$

and substituting it into Eq. 89, along with the square root of Eq. 85,

$$\tilde{r}_p = \frac{2\tilde{\varepsilon}_\circ\, \tilde{\alpha}^3\sqrt{1-\beta^2}}{\beta^3}$$

yields

(90) $$\tilde{G} = \frac{8\tilde{\alpha}^5\tilde{e}^2c^2\tilde{\varepsilon}_\circ\left(1-\beta^2\right)^{3/2}}{3\pi\, \tilde{m}_e\beta^3}$$

Taking Eq. 62 in the following form,

$$\frac{\tilde{e}}{\tilde{m}_e} = \frac{\pi\beta}{2\tilde{\mu}_\circ^2}$$

and substituting it into Eq. 90 with the known identity $\tilde{\varepsilon}_\circ = 1/\tilde{\mu}_\circ c^2$ from Maxwell relation in its initial form yields

(91) $$\tilde{G} = \frac{4\tilde{\alpha}^5\left(1-\beta^2\right)^{3/2}\tilde{e}}{3\beta^2\tilde{\mu}_\circ^3}$$

Observing Eq. 91 with Eq. 88 here

$$\tilde{G} = \left(2\pi^2\tilde{\alpha}^8\beta^4\sqrt{1-\beta^2}\right)\left(\frac{8\pi^5\sqrt{1-\beta^2}}{\beta^6c^2\tilde{\mu}_\circ^3}\right)$$

It is now possible to deduce the following insights from the last equation:

a. The component $\tilde{\mu}_\circ^3$ is found in the denominator of both equations.

b. Eq. 91 contains the component $\widetilde{e}$ in its numerator, and therefore it is possible to assume that the component $\widetilde{e}$ is *hidden* in the Eq. 88 numerator.

Based on the expected initial value of the elementary charge $\widetilde{e}$, as calculated by Eq. 61:

$$\widetilde{e} = 1.604332974 \times 10^{-19} [C]$$

and the conclusion from the above insights regarding Eq. 87 and Eq. 91, I realized that the following expression for $\widetilde{e}$, which accurately matches the expression in Eq. 61, is the first expression in the parenthesis on the right side of Eq. 88, here

(92) $$\widetilde{e} = 2\pi^2 \widetilde{\alpha}^8 \beta^4 \sqrt{1-\beta^2}$$

Note that the elementary charge $\widetilde{e}$ in Eq. 92 is based on theoretical expressions $\widetilde{\alpha}$ β and π only!

Substituting the following data values into Eq. 92

Key constant: $\beta = 0.178145016$

Initial fine structure constant: $\widetilde{\alpha} = 7.315334297 \times 10^{-3}$

Yields the theoretical initial value of the elementary charge $\widetilde{e}$:

$$\widetilde{e} = 1.60433419 \times 10^{-19} [C]$$

This value can be compared to the expected initial value, which is based on the experimentally received value of the speed of light in vacuum c:

$$\widetilde{e} = 1.604332974 \times 10^{-19} [C]$$

The form of the expression formulated here for $\widetilde{e}$ is not trivial!

One can see that only the specific combination of components in Eq. 87 gives the theoretical initial value, which matches the expected one. Placing the theoretical initial value received for $\widetilde{e}$ from Eq. 92 fulfills the assumption made in Eq. 61, here

$$\frac{\widetilde{e} c^2 \beta}{48 \widetilde{\alpha}^2} = 1$$

To show the development of the expression calculating the theoretical value of the speed of light in vacuum c which matches the experimentally received value, please observe Eq. 88 and Eq. 91. It is noticeable that the component c^2 appears in the denominator of Eq. 88, leading to the assumption that it is hidden in Eq. 91 as well.

Equating Eq. 88 with Eq. 91, yields

(93) $$\frac{4\widetilde{\alpha}^5 (1-\beta^2)^{3/2} \widetilde{e}}{3\beta^2 \widetilde{\mu}_\circ^3} = \left(2\pi^2 \widetilde{\alpha}^8 \beta^4 \sqrt{1-\beta^2}\right)\left(\frac{8\pi^5 \sqrt{1-\beta^2}}{\beta^6 c^2 \widetilde{\mu}_\circ^3}\right)$$

Reducing the $\widetilde{\mu}_\circ^3$ and $\widetilde{e}$ components, using the substitution of Eq. 92, yields

(94) $$\frac{4\widetilde{\alpha}^5 (1-\beta^2)^{3/2}}{3\beta^2} = \frac{8\pi^5 \sqrt{1-\beta^2}}{\beta^6 c^2}$$

Isolating the speed of light in vacuum c and reducing Eq. 94 gives

$$(95)\qquad c^2 = \frac{6\pi^5}{\beta^4\tilde{\alpha}^5\left(1-\beta^2\right)}$$

Placing the relation $\left(1-\beta^2\right)=\pi^6\beta^4$ from Eq. 38 into Eq. 95 yields

$$(96)\qquad c^2 = \frac{6}{\pi\,\tilde{\alpha}^5\beta^8}$$

Eq. 96 allows the calculation of the speed of light in vacuum c on a theoretical basis, using the components $\tilde{\alpha}$, π and β only!

Substituting the following data values into Eq. 96

Key constant: $\beta = 0.178145016$

Initial fine structure constant: $\tilde{\alpha} = 7.315334297\times10^{-3}$

Yields the theoretical value of the speed of light in vacuum c:

$c = 2.99792344\times10^{8}\left[m/s\right]$

For comparison, the experimentally received value [2] of the speed of light in vacuum c is

$c = 2.99792458\times10^{8}\left[m/s\right]$

As it can be seen, the deviation of the theoretical value with relation to the experimentally received value is 114 meters, for a value measured in thousands of kilometers!

Substituting the theoretical value for c into Eq. 83

$$\tilde{\mu}_{\circ}^3 = \frac{1}{c^2}\left(\frac{\beta}{\sqrt{1-\tilde{\alpha}^2/\beta^2}}\right)$$

Yields the theoretical initial value of the *permeability of vacuum constant* $\tilde{\mu}_{\circ}$,

$\tilde{\mu}_{\circ} = 1.256513691\times10^{-6}\left[N\;A^{-2}\right]$

Substituting the theoretical initial values of the elementary charge $\tilde{e}$, $\tilde{\alpha}$, π, and β into Eq. 84

$$\tilde{\varepsilon}_{\circ} = \left[4\pi\,\tilde{e}\sqrt{1-\beta^2}\right]^{2/3}\left(\frac{\sqrt{1-\tilde{\alpha}^2/\beta^2}}{\beta}\right)$$

Yields the theoretical initial value of the *permittivity of vacuum constant* $\tilde{\varepsilon}_{\circ}$:

$\tilde{\varepsilon}_{\circ} = 8.855063892\times10^{-12}\left[F\;m^{-1}\right]$

Substituting the theoretical initial value of the permittivity of vacuum constant $\tilde{\varepsilon}_{\circ}$ in the relation $\tilde{a}_e = 6\tilde{\varepsilon}_{\circ}$ from Eq. 15 gives the theoretical initial value of the *Bohr radius* $\tilde{a}_e$:

$\widetilde{a}_e = 5.313038333 \times 10^{-11} [m]$

Substituting the theoretical initial values found for $\widetilde{e}$ and $\widetilde{\mu}_\circ$ into Eq. 62

$$\widetilde{m}_e = \frac{2\widetilde{e}\widetilde{\mu}_\circ^2}{\pi\beta}$$

Yields the theoretical initial value of the *electron mass* $\widetilde{m}_e$

$\widetilde{m}_e = 9.051816380 \times 10^{-31} [kg]$

Substituting the theoretical initial value of the *electron mass* $\widetilde{m}_e$, into Eq. 32

$$\widetilde{m}_p = \frac{3\pi\,\widetilde{m}_e}{4\widetilde{\alpha}\beta\sqrt{1-\beta^2}}$$

Along with the initial value of the fine structure constant $\widetilde{\alpha}$, the key constant β, and π, yields the theoretical initial value of the *proton mass* $\widetilde{m}_p$:

$\widetilde{m}_p = 1.663190691 \times 10^{-27} [kg]$

Substituting the theoretical initial value of the *Bohr radius* $\widetilde{a}_e$, along with the fine structure constant $\widetilde{\alpha}$, the key constant β, and π, into Eq. 33

$$\widetilde{r}_p = \widetilde{a}_e \frac{4\widetilde{\alpha}^2}{3\pi}$$

Yields the theoretical initial value of the *proton radius* $\widetilde{r}_p$:

$\widetilde{r}_p = 1.206702376 \times 10^{-15} [m]$

For comparison, the *estimated value* for the proton radius, based on experiments, as was discussed in the first chapter, Eq. 2, is

$r_p = R_\circ \approx 1.2 \times 10^{-15} [m]$

6
The η constant

The η constant represents the ratio between the initial value of the permeability of vacuum $\tilde{\mu}_\circ$ and the experimentally received value $\mu_\circ$, and the ratio between the experimentally received value of the permittivity of vacuum $\varepsilon_\circ$ and its initial value $\tilde{\varepsilon}_\circ$:

$$\eta = \frac{\tilde{\mu}_\circ}{\mu_\circ} = \frac{\varepsilon_\circ}{\tilde{\varepsilon}_\circ}$$

This ratio has a significant part in the continuation of the theory development. The ratio is identical for both constants as a result of the Maxwell identity[4] $\varepsilon_\circ = 1/\mu_\circ c^2$.

Substituting the following data values into the above ratios
Theoretical initial value of permeability of vacuum constant: $\tilde{\mu}_\circ = 1.25651369 \times 10^{-6} \left[N\ A^{-2}\right]$
Known value which matches the experiments[2]: $\mu_\circ = 1.256637061 \times 10^{-6} \left[N\ A^{-2}\right]$
Theoretical initial value of permittivity of vacuum constant: $\tilde{\varepsilon}_\circ = 8.85506389 \times 10^{-12} \left[F\ m^{-1}\right]$
Known value which matches the experiments[2]: $\varepsilon_\circ = 8.854187818 \times 10^{-12} \left[F\ m^{-1}\right]$

Yields the expected value of the η constant:

$$\eta = \frac{\tilde{\mu}_\circ}{\mu_\circ} = \frac{\varepsilon_\circ}{\tilde{\varepsilon}_\circ} = 0.999901825$$

This value is an estimation only, since it is based on the known experimental values of $\mu_\circ$ and $\varepsilon_\circ$.
A final value will be based on theoretical values only as will be shown later on.

PART 3
The Final Values of the Fundamental Constants

1
The Proton/Electron interaction influence On the initial constants values

Entering the proton/electron interaction factors in the hydrogen atom into the initial values expressions corrects the initial values found for the different constants to their final values, which match the experiments. These interaction factors are:

a. The reduced electron mass factor: It is derived from the proton/electron movement around their common mass center in the hydrogen atom. This factor in its exact value always contains the initial proton and the initial electron masses!

(97) $$(1 - \tilde{m}_e / \tilde{m}_p)$$

b. The electron mass under the influence of the reduced electron mass factor**:** Represents the influence of the reduced mass factor on the initial electron mass.

(98) $$\tilde{m}_e(1 - \tilde{m}_e / \tilde{m}_p)$$

c. The Bohr radius under the influence of the reduced electron mass factor: Represents the influence of the reduced mass factor on the Bohr radius.

(99) $$\frac{\tilde{a}_e}{(1 - \tilde{m}_e / \tilde{m}_p)}$$

d. The γ Lorentz factor of proton/electron velocities interaction**:** Arises from deriving a relative velocity express by the interaction between the proton's spinning velocity and the electron's velocity at the Bohr radius. Placing the relation $\tilde{v}_e = c\tilde{\alpha}$ from Eq. 6 and $v_p = c\beta$ from Eq. 7 forms:

(100) $$\sqrt{1 - \tilde{v}_e^2/v_p^2} = \sqrt{1 - \tilde{\alpha}^2/\beta^2}$$

The effective value of the γ factor is calculated by the expression mentioned at Eq. 81 shown here:

$$\gamma_{proton/electron} = \frac{1}{192\pi^4\tilde{\alpha}^2} \equiv \sqrt{1 - \tilde{\alpha}^2/\beta^2}$$

It is important to note that calculating of the γ factor value is done according to the term $\frac{1}{192\pi^4\tilde{\alpha}^2}$ from Eq. 81 using only the $\tilde{\alpha}$ constant and π throughout the entire development although in fact this term $\sqrt{1-\tilde{\alpha}^2/\beta^2}$ will appear in the equations from now on.

e. The relative electron velocity at the Bohr radius under the influence of the γ factor:

(101) $$\frac{\tilde{v}_e}{\sqrt{1-\tilde{\alpha}^2/\beta^2}}$$

f. The correction factor μ_e/μ_B related to the electron magnetic moment μ_e.
It is possible to present the electron kinetic energy at the Bohr radius E_k with the influence of the reduced mass factor from Eq. 98, and with the influence of the γ factor on the electron's velocity from Eq. 101 to form:

$$E_k = \frac{1}{2} \underbrace{\tilde{m}_e(1-\tilde{m}_e/\tilde{m}_p)}_{\text{Reduced electron mass}} \left[\underbrace{\frac{\tilde{v}_e^2}{(1-\tilde{\alpha}^2/\beta^2)}}_{\text{Relative electron velocity}} \right]$$

Or

$$(102) \qquad E_k = \frac{1}{2}\tilde{m}_e\tilde{v}_e^2 \left[\frac{(1-\tilde{m}_e/\tilde{m}_p)}{(1-\tilde{\alpha}^2/\beta^2)} \right]$$

***The electron magnetic moment to Bohr magneton ratio** μ_e/μ_B is the terms in the brackets on the right flank of Eq. 102: (please refer to page 88 for a detailed process on how it is obtained!)

$$(103) \qquad \mu_e/\mu_B = \frac{(1-\tilde{m}_e/\tilde{m}_p)}{(1-\tilde{\alpha}^2/\beta^2)}$$

Substituting the following data values in Eq. 103:
Effective value of the γ factor: $\sqrt{1-\tilde{\alpha}^2/\beta^2} = \frac{1}{192\pi^4\tilde{\alpha}^2} = 0.999150536$
Reduced mass factor: $(1-\tilde{m}_e/\tilde{m}_p) = 0.999455756$
Initial value of the electron mass: $\tilde{m}_e = 9.051816380 \times 10^{-31}[kg]$
Initial value of the proton mass: $\tilde{m}_p = 1.663190691 \times 10^{-27}[kg]$
Initial value of the fine structure constant: $\tilde{\alpha} = 7.315334297 \times 10^{-3}$
Yields the electron magnetic moment to Bohr magneton ratio (μ_e/μ_B):
$\mu_e/\mu_B = 1.001155925$
For comparison, the experimentally calculated value of the factor μ_e/μ_B is the following (please see pages 90 and 91 for explanation of why is there a slight difference from the theoretical one!):
$\mu_e/\mu_B = 1.001159653$

2
The Expressions for the Final Values of the Fundamental Constants

This chapter shows the development that yields the theoretical expressions calculating the final values of the different constants. These final values *match the experimental values*!

2-1. The theoretical expression for the elementary charge e:

Starting with Eq. 80, the initial value expression for the permeability of vacuum constant

$\tilde{\mu}_{\circ}^{3} = 4\pi\tilde{e}\sqrt{1-\beta^{2}}$

applying the known identity $\tilde{\mu}_{\circ} = 1/\tilde{\varepsilon}_{\circ}c^{2}$ from Maxwell relations in its initial form to Eq. 80 yields

(104) $$\left[\frac{1}{\tilde{\varepsilon}_{\circ}c^{2}}\right]^{3} = 4\pi\tilde{e}\sqrt{1-\beta^{2}}$$

Taking the Maxwell identity[4] presented by the known experimental values for the permeability of vacuum constant $\mu_{\circ}$ and the permittivity of vacuum constant $\varepsilon_{\circ}$

$c^{2} = 1/\varepsilon_{\circ}\mu_{\circ}$,

and replacing this identity into the denominator of the left term in Eq. 104 yields

$$\left[\frac{\varepsilon_{\circ}\mu_{\circ}}{\tilde{\varepsilon}_{\circ}}\right]^{3} = 4\pi\tilde{e}\sqrt{1-\beta^{2}}$$

Or, by multiplying and dividing nominator and denominator on the left side of the equation by **6**,

(105) $$\left[\frac{6\varepsilon_{\circ}\mu_{\circ}}{6\tilde{\varepsilon}_{\circ}}\right]^{3} = 4\pi\tilde{e}\sqrt{1-\beta^{2}}$$

and substituting Eq. 15

$$\left(\frac{\tilde{a}_{e}}{6\tilde{\varepsilon}_{\circ}}\right) = 1 \rightarrow \tilde{a}_{e} = 6\tilde{\varepsilon}_{\circ}$$

into Eq. 99, which presents the Bohr radius under the influence of the reduced mass factor:-

$$\frac{\tilde{a}_{e}}{(1-\tilde{m}_{e}/\tilde{m}_{p})}$$

yields

(106) $$\frac{6\tilde{\varepsilon}_\circ}{(1-\tilde{m}_e/\tilde{m}_p)}$$

In order to get the expression of Eq. 106 on the left side of Eq. 105, both sides of Eq. 105 are multiplied by the reduced mass factor in cube $(1-\tilde{m}_e/\tilde{m}_p)^3$

$$\left[\frac{6\varepsilon_\circ\mu_\circ(1-\tilde{m}_e/\tilde{m}_p)}{6\tilde{\varepsilon}_\circ}\right]^3 = 4\pi\,\tilde{e}\sqrt{1-\beta^2}\,(1-\tilde{m}_e/\tilde{m}_p)^3$$

Or, by placing the relation $\eta = \frac{\varepsilon_\circ}{\tilde{\varepsilon}_\circ}$, we obtain

(107) $$\eta^3\mu_\circ^3(1-\tilde{m}_e/\tilde{m}_p)^3 = 4\pi\,\tilde{e}\sqrt{1-\beta^2}\,(1-\tilde{m}_e/\tilde{m}_p)^3$$

Dividing both sides by (η^3) and rearranging yields

(108) $$\mu_\circ^3(1-\tilde{m}_e/\tilde{m}_p)^3 = 4\pi\sqrt{1-\beta^2}\left[\tilde{e}\left(\frac{(1-\tilde{m}_e/\tilde{m}_p)^3}{\eta^3}\right)\right]$$

Eq. 108 has a similar format to Eq. 80. It differs by incorporating the interaction influence factors on the initial values. While Eq. 80 is expressing the relation between the initial $\tilde{e}$ versus $\tilde{\mu}_\circ$ values, Eq. 108 reflects the relation between these constants received at the experiments. On the left side of Eq. 108 is $\mu_\circ$, which suggests that the experimental matching of the elementary charge e will be the term in the brackets on the right side:

(109) $$e = \tilde{e}\left(\frac{(1-\tilde{m}_e/\tilde{m}_p)^3}{\eta^3}\right)$$

To check the justification, we'll substitute the following data values into Eq. 109:

Theoretical reduced electron mass factor: $(1-\tilde{m}_e/\tilde{m}_p) = 0.999455756$

Expected value of the η constant: $\eta = \frac{\tilde{\mu}_\circ}{\mu_\circ} = 0.999901825$

Theoretical initial value of the elementary charge: $\tilde{e} = 1.60433419\times10^{-19}[C]$

Yields the theoretical value of the elementary charge e, which matches the experimentally received value:

$$e = 1.602188\times10^{-19}[C]$$

Hence, placing e into Eq. 108 on the right side instead of the brackets yields

(110) $$\mu_\circ^3(1-\tilde{m}_e/\tilde{m}_p)^3 = 4\pi e\sqrt{1-\beta^2}$$

Or, isolating the elementary charge e component yields

(111) $$e = \frac{\mu_\circ^3 (1-\tilde{m}_e/\tilde{m}_p)^3}{4\pi\sqrt{1-\beta^2}}$$

2-2. The theoretical expression for the fine structure constant α:

Multiplying both sides of Eq. 16 in its initial form by the reduced mass factor $(1-\tilde{m}_e/\tilde{m}_p)$, and placing

$$\tilde{v}_e = c\tilde{\alpha}$$

and

$$\tilde{e}^2 = \frac{\tilde{m}_e \tilde{v}_e^2 4\pi\, \tilde{a}_e^2}{6}$$

yields

(112) $$\tilde{e}^2(1-\tilde{m}_e/\tilde{m}_p) = \frac{\tilde{m}_e(1-\tilde{m}_e/\tilde{m}_p)\, c^2\tilde{\alpha}^2 4\pi\tilde{a}_e^2}{6}$$

We see that the electron reduced mass factor $(1-\tilde{m}_e/\tilde{m}_p)$ is multiplying $\tilde{m}_e$ at the numerator on the right side of Eq. 112**,** as it should be. The Bohr radius $\tilde{a}_e$ is affected also by the reduced mass factor. To reflect this, both sides of Eq. 112 are divided by $(1-\tilde{m}_e/\tilde{m}_p)^2$:

(113) $$\tilde{e}^2(1-\tilde{m}_e/\tilde{m}_p)\left[\frac{1}{(1-\tilde{m}_e/\tilde{m}_p)^2}\right] = \frac{\tilde{m}_e(1-\tilde{m}_e/\tilde{m}_p)\, c^2\tilde{\alpha}^2 4\pi\, \tilde{a}_e^2}{6(1-\tilde{m}_e/\tilde{m}_p)^2}$$

Now, in order to generate the expression of $\tilde{v}_e$ with the γ factor as in Eq. 101, both sides of Eq. 113 are divided by $(1-\tilde{\alpha}^2/\beta^2)$ because of $\tilde{v}_e^2$ (the initial electron velocity raised to power of **2**), and after rearrangement, yields

(114) $$\tilde{e}^2\left[\frac{(1-\tilde{m}_e/\tilde{m}_p)}{(1-\tilde{\alpha}^2/\beta^2)}\right]\left[\frac{1}{(1-\tilde{m}_e/\tilde{m}_p)^2}\right] = \left[\tilde{m}_e(1-\tilde{m}_e/\tilde{m}_p)\right]\left[\frac{c^2\tilde{\alpha}^2}{(1-\tilde{\alpha}^2/\beta^2)}\right]\frac{4\pi}{6}\left[\frac{\tilde{a}_e^2}{(1-\tilde{m}_e/\tilde{m}_p)^2}\right]$$

Eq. 109 can now be presented as

$$\tilde{e} = e\left[\frac{\eta^3}{(1-\tilde{m}_e/\tilde{m}_p)^3}\right]$$

Substituting Eq. 109 into the left term of Eq. 114**,** yields

(115)

$$e^2\left[\frac{\eta^3}{(1-\tilde{m}_e/\tilde{m}_p)^3}\right]^2\left[\frac{(1-\tilde{m}_e/\tilde{m}_p)}{(1-\tilde{\alpha}^2/\beta^2)}\right]\left[\frac{1}{(1-\tilde{m}_e/\tilde{m}_p)^2}\right] = \left[\tilde{m}_e(1-\tilde{m}_e/\tilde{m}_p)\right]\left[\frac{c^2\tilde{\alpha}^2}{(1-\tilde{\alpha}^2/\beta^2)}\right]\frac{4\pi}{6}\left[\frac{\tilde{a}_e^2}{(1-\tilde{m}_e/\tilde{m}_p)^2}\right]$$

Reducing as needed and rearranging Eq. 115 yields

$$e^2\left[\frac{\eta^6}{(1-\tilde{m}_e/\tilde{m}_p)^7\left(1-\tilde{\alpha}^2/\beta^2\right)}\right]=\frac{\tilde{m}_e c^2\tilde{\alpha}^2 4\pi\tilde{a}_e^2}{6(1-\tilde{m}_e/\tilde{m}_p)\left(1-\tilde{\alpha}^2/\beta^2\right)}$$

or by isolating the elementary charge e,

(116) $$e^2=\frac{\tilde{m}_e c^2 4\pi\,\tilde{a}_e^2}{6\left(1-\tilde{m}_e/\tilde{m}_p\right)\left(1-\tilde{\alpha}^2/\beta^2\right)}\left[\tilde{\alpha}^2\left(\frac{\left(1-\tilde{m}_e/\tilde{m}_p\right)^7\left(1-\tilde{\alpha}^2/\beta^2\right)}{\eta^6}\right)\right]$$

Eq. 116 has a similar format to Eq. 16. It differs by incorporating the interaction influence factors on the initial values .While Eq. 16 expresses the relation between the initial $\tilde{\alpha}^2$ versus $\tilde{e}^2$ values, Eq. 116 reflects the relation between these constants received at the experiments. On the left side of Eq. 116 we have e^2, which suggests that α^2 will be the term in the square brackets on the right side.

$$\alpha^2=\tilde{\alpha}^2\left(\frac{(1-\tilde{m}_e/\tilde{m}_p)^7\left(1-\tilde{\alpha}^2/\beta^2\right)}{\eta^6}\right)$$

After extracting the square root, we obtain the theoretical expression of the fine strucuture constant

(117) $$\alpha=\tilde{\alpha}\frac{\sqrt{1-\tilde{\alpha}^2/\beta^2}\times(1-\tilde{m}_e/\tilde{m}_p)^{7/2}}{\eta^3}$$

To check the justification, substituting the following data values into Eq. 117

Theoretical value of the reduced mass factor: $\left(1-\tilde{m}_e/\tilde{m}_p\right)=0.999455756$

Initial value of the fine structure constant: $\tilde{\alpha}=7.315334297\times10^{-3}$

Expected value of the η constant: $\eta=\dfrac{\tilde{\mu}_\circ}{\mu_\circ}=0.999901825$

γ factor: $\left(1-\tilde{\alpha}^2/\beta^2\right)=(0.99915056)^2$

Yields

$\alpha=7.29735588\times10^{-3}$

Or, in its more familiar inverse form:

$\alpha^{-1}=137.0359368$

As a reminder**,** the calculation of α here is based on the expected value of the η constant! The calculation of the exact value of the fine structure constant α by Eq. 117 will be done once the η constant is theoretically determined later on.

Following the last finding, we substitute α into Eq. 116 on the right side, instead of the square brackets, as follows

(118) $$e^2=\frac{\tilde{m}_e c^2\alpha^2 4\pi\tilde{a}_e^2}{6\left(1-\tilde{m}_e/\tilde{m}_p\right)\left(1-\tilde{\alpha}^2/\beta^2\right)}$$

The e^2 in Eq. 118 which includes the interaction factors, *is identical and equal* to Eq.16

$e^2 = m_e v_e^2 4\pi\, a_e \varepsilon_\circ$

Eq. 118 represents the expression for calculating the elementary charge e value that matches the experimental one!

2-3. Additional useful relations between the elementary charge e and the fine structure constant α:

Take Eq.16 in its initial form

(119) $$\tilde{e}^2 = \frac{\tilde{m}_e \tilde{v}_e^2 4\pi\, \tilde{a}_e^2}{6}$$

and place $\tilde{v}_e = c\tilde{\alpha}$, to yield

(120) $$\tilde{e}^2 = \frac{\tilde{m}_e c^2\, \tilde{\alpha}^2 4\pi\, \tilde{a}_e^2}{2\times 3}$$

Dividing Eq. 118 by Eq. 120 yields the following expression, which will be useful later on!

$$\frac{e^2}{\tilde{e}^2} = \left(\frac{\alpha^2}{\tilde{\alpha}^2}\right)\frac{1}{\left(1-\tilde{m}_e/\tilde{m}_p\right)\left(1-\tilde{\alpha}^2/\beta^2\right)}$$

Or, after extracting the square root,

(121) $$\frac{e}{\tilde{e}} = \left(\frac{\alpha}{\tilde{\alpha}}\right)\frac{1}{\sqrt{1-\tilde{m}_e/\tilde{m}_p}\,\sqrt{1-\tilde{\alpha}^2/\beta^2}}$$

2-4. The **speed of light in the atomic domain** c_n**:**

This section will complete the explanation on why the speed of light in vacuum c is excluded from the initial values rule.

Substituting the following data values into Eq. 82, here

$$c^2 = \frac{192\pi^4\tilde{\alpha}^2\beta}{\tilde{\mu}_\circ^3} = \frac{\beta}{\tilde{\mu}_\circ^3\sqrt{1-\tilde{\alpha}^2/\beta^2}}$$

Initial value of the fine structure constant: $\tilde{\alpha} = 7.315334297\times 10^{-3}$

Initial value of the permeability of vacuum constant: $\tilde{\mu}_\circ = 1.256513691\times 10^{-6}\left[N\ A^{-2}\right]$

Key constant: $\beta = 0.178145016$

Yields the same theoretical value for the speed of light in vacuum c as was calculated by Eq. 96:

$$c = 2.99792344\times 10^8\left[m\ s^{-1}\right]$$

Based on the change of $\tilde{\mu}_\circ$ from Eq. 80 to Eq. 110, done in order to find the value of the elementary charge e, a similar change is made for $\tilde{\mu}_\circ$ at Eq. 82, where $\tilde{\mu}_\circ$ turns to $\mu_\circ$.

Note that the speed of light in Eq. 82 must change as well, marked as c_n.

Placing the expression $\mu_\circ^3(1-\tilde{m}_e/\tilde{m}_p)^3$ instead of $\tilde{\mu}_\circ^3$ in the denominator of Eq. 82 yields

(122) $$c_n^2 = \frac{\beta}{\mu_\circ^3 (1 - \tilde{m}_e / \tilde{m}_p)^3 \sqrt{1 - \tilde{\alpha}^2/\beta^2}}$$

Substituting the following data values into Eq. 122

Initial value of the fine structure constant: $\tilde{\alpha} = 7.315334297 \times 10^{-3}$
Theoretical initial value of the electron mass: $\tilde{m}_e = 9.051816380 \times 10^{-31} [kg]$
Theoretical initial value of the proton mass: $\tilde{m}_p = 1.663190691 \times 10^{-27} [kg]$
Key constant: $\beta = 0.178145016$
Experimentally received value of the permeability of vacuum constant[2]:
$\mu_\circ = 1.26637061 \times 10^{-6} [N\ A^{-2}]$
Yields

$$c_n = 2.999930699 \times 10^8 [m\ s^{-1}]$$

This leads us to the following conclusion: ***The speed of light is higher than*** c ***in the atomic domain***! As related to the experiment based on the EPR paradox,[7] this conclusion can explain the communication between the two particles in a speed that exceeds the speed of light in vacuum c. Feynman[6] suggested that possibility as well in his book, *QED*, claiming that light can travel in a higher speed than the speed of light in vacuum c, and considered this possibility in his own calculations!

The possibility of the existence of a higher speed of light in the atomic domain c_n, in comparison with the known speed of light c, yields the following conclusions:
The higher speed of light allows the photons to *possess mass* without contradicting the laws of the special theory of relativity. This means that in the following expression,[4] the Lorentz γ factor found in the denominator never reaches an infinite value as the photon travels at the speed of light in vacuum c. It also means that **this theory presents a nonlocal Hidden Variables**!

That's because the expression $\left(\frac{c^2}{c_n^2}\right)$ at the denominator is always smaller than **1**!

$$\frac{m_{photon}}{\sqrt{1 - c^2/c_n^2}} \neq \infty$$

The *uncertainty principle of Heisenberg,*[7] formulated as inequality and called *Heisenberg inequality*

$$\Delta x \times \Delta(mv) \geq \hbar$$

$$\hbar = \frac{h}{2\pi}$$

Thus, if the speed of light in vacuum is affected by the interaction factors, then the Planck constant is also affected

To check, divide Eq. 16,

$e^2 = m_e v_e^2 4\pi\, a_e \varepsilon_\circ$

by Eq. 17a,

$h = m_e v_e 2\pi a_e$

and substitute $v_e = c\alpha$ and $\varepsilon_\circ = \dfrac{1}{\mu_\circ c^2}$ to obtain for h

$$h = \frac{e^2 \mu_\circ c}{2\alpha}$$

Now, e, $\mu_\circ$ and α stands for values that match the experimentaly received values and that were obtained from involvement of the interaction factors.

Substituting c with c_n yields a higher value for the Planck constant!

$$h_n = \frac{e^2 \mu_\circ c_n}{2\alpha}$$

Where h_n is the Planck constant in the atomic domain!
Now, to get an idea of the magnitude of h_n, substitute the following data values in it:
Experimentally received value of the elementary charge:[2] $e = 1.6021892 \times 10^{-19}[C]$
Speed of light in the atomic domain received theoreticaly: $c_n = 2.999930699 \times 10^8 [m\ s^{-1}]$
Experimentally received value of the fine structure constant: $\alpha^{-1} = 137.036$

Yields the Planck constant in the atomic domain h_n

$$h_n = 6.630595108 \times 10^{-34} [J\ s]$$

This value is higher than the known Planck constant value due to c_n. Now, placing $\hbar_n$ instead of $\hbar$ on the right side of Heisenberg inequality suggests that at certain conditions of the particle location and angular momentum, the inequality sign can be inverted due to the higher value of the Planck constant in the atomic domain!

$$\Delta x \times \Delta(mv) \leq \hbar_n$$

This gives way to the possibility for the **sub-atomic particles to violate the uncertainty principle**!

2-5. The **speed of light in the atomic domain** c_n in relation to other constants**:**

It is possible to present another expression for the known value of the permittivity of vacuum constant $\varepsilon_\circ$ that includes the speed of light in the atomic domain c_n:

(123) $$\left(\frac{c^2}{c_n^2}\right)\frac{\varepsilon_\circ}{\left(1 - \tilde{m}_e / \tilde{m}_p\right)} = \left[4\pi\, e\sqrt{1-\beta^2}\right]^{2/3} \frac{\sqrt{1-\tilde{\alpha}^2/\beta^2}}{\beta}$$

The equality in Eq.123 can be verified by substituting c_n from Eq. 122, along with the known identity $\varepsilon_\circ c^2 = 1/\mu_\circ$ from Maxwell relations, and extracting the square root and raising both sides to a power of **3**, resulting in Eq. 110, shown here

$$\mu_\circ^3 (1 - \widetilde{m}_e / \widetilde{m}_p)^3 = 4\pi e\sqrt{1-\beta^2}$$

In addition, substituting Eq. 80 here

$$\widetilde{\mu}_\circ^3 = 4\pi\, \widetilde{e}\sqrt{1-\beta^2}$$

into Eq. 82 here

$$c^2 = \frac{\beta}{\widetilde{\mu}_\circ^3 \sqrt{1 - \widetilde{\alpha}^2/\beta^2}}$$

yields

$$(124) \qquad c^2 = \frac{\beta}{4\pi\, \widetilde{e}\sqrt{1-\beta^2}\sqrt{1-\widetilde{\alpha}^2/\beta^2}}$$

Substituting Eq. 110 here

$$\mu_\circ^3 (1 - \widetilde{m}_e / \widetilde{m}_p)^3 = 4\pi\, e\sqrt{1-\beta^2}$$

into Eq. 122 here

$$c_n^2 = \frac{\beta}{\mu_\circ^3 (1 - \widetilde{m}_e / \widetilde{m}_p)^3 \sqrt{1 - \widetilde{\alpha}^2/\beta^2}}$$

yields

$$(125) \qquad c_n^2 = \frac{\beta}{4\pi\, e\sqrt{1-\beta^2}\sqrt{1-\widetilde{\alpha}^2/\beta^2}}$$

Dividing Eq. 124 by Eq. 125 yields

$$(126) \qquad \frac{c^2}{c_n^2} = \frac{e}{\widetilde{e}}$$

Substituting the square root of Eq. 121 here

$$\frac{e}{\widetilde{e}} = \left(\frac{\alpha}{\widetilde{\alpha}}\right) \frac{1}{\sqrt{1 - \widetilde{m}_e / \widetilde{m}_p}\sqrt{1 - \widetilde{\alpha}^2 / \beta^2}}$$

into Eq. 126 yields a relation which will be useful in further developments:

(127) $$\frac{c^2}{c_n^2}=\left(\frac{\alpha}{\tilde{\alpha}}\right)\frac{1}{\sqrt{1-\tilde{m}_e/\tilde{m}_p}\sqrt{1-\tilde{\alpha}^2/\beta^2}}$$

2-6. The theoretical expressions for the electron mass m_e and the Bohr radius a_e:
Start by equalizing Eq. 118 with Eq. 16**,** as shown here

$$\frac{\tilde{m}_e v_e^2 4\pi\tilde{a}_e^2}{6(1-\tilde{\alpha}^2/\beta^2)(1-\tilde{m}_e/\tilde{m}_p)}=m_e v_e^2 4\pi\ a_e\varepsilon_\circ$$

Multiplying both sides by of the equality by $\left(\frac{\tilde{\alpha}^2}{\alpha^2}\right)\frac{6(1-\tilde{m}_e/\tilde{m}_p)^2}{4\pi m_e v_e^2\tilde{a}_e^2}$ in order to transpose terms, and substituting **one** of the Bohr radius components with the relation $\tilde{a}_e=6\tilde{\varepsilon}_\circ$ from Eq. 15 yields

(128) $$\tilde{m}_e(1-\tilde{m}_e/\tilde{m}_p)\left[\frac{\tilde{v}_e^2}{(1-\tilde{\alpha}^2/\beta^2)}\right]\frac{1}{m_e v_e^2}=\left(\frac{a_e}{\tilde{a}_e}\right)\left(\frac{\varepsilon_\circ}{\tilde{\varepsilon}_\circ}\right)\left(\frac{\tilde{\alpha}^2}{\alpha^2}\right)(1-\tilde{m}_e/\tilde{m}_p)^2$$

Note that (v_e^2) in the numerator of the left term of Eq. 118 becomes $(\tilde{v}_e^2)$ in Eq. 128, as part of components reduction!
Multiplying both sides of Eq. 128 by the right side of the equality, while substituting the relation

$\frac{\varepsilon_\circ}{\tilde{\varepsilon}_\circ}=\eta$, yields

(129) $$\left[\frac{\tilde{m}_e(1-\tilde{m}_e/\tilde{m}_p)^3}{\tilde{a}_e}\right]\left[\frac{\tilde{v}_e^2}{(1-\tilde{\alpha}^2/\beta^2)}\right]\left(\frac{a_e}{m_e v_e^2}\right)\left(\eta\times\frac{\tilde{\alpha}^2}{\alpha^2}\right)=\left(\frac{a_e^2}{\tilde{a}_e^2}\right)\left(\eta^2\times\frac{\tilde{\alpha}^4}{\alpha^4}\right)(1-\tilde{m}_e/\tilde{m}_p)^4$$

Dividing both sides of the equality of Eq. 129 by $(1-\tilde{m}_e/\tilde{m}_p)$ and rearranging yields

(130) $$\frac{\tilde{m}_e(1-\tilde{m}_e/\tilde{m}_p)^2}{m_e v_e^2}\left[\frac{\tilde{v}_e^2}{(1-\tilde{\alpha}^2/\beta^2)}\right]\left(\frac{a_e}{\tilde{a}_e}\right)\left(\eta\times\frac{\tilde{\alpha}^2}{\alpha^2}\right)=\left(\frac{a_e^2}{\tilde{a}_e^2}\right)(1-\tilde{m}_e/\tilde{m}_p)^3\left(\eta^2\times\frac{\tilde{\alpha}^4}{\alpha^4}\right)$$

Defining the right side of Eq. 130 as equal to **1**, as here

$$\left(\frac{a_e^2}{\tilde{a}_e^2}\right)(1-\tilde{m}_e/\tilde{m}_p)^3\left(\eta^2\times\frac{\tilde{\alpha}^4}{\alpha^4}\right)=1$$

allows for isolating an expression that includes the ratio $a_e/\tilde{a}_e$, from which it is possible to further isolate the Bohr radius a_e that matches the experiments.
Substituting the following data values

Initial value of the fine structure constant: $\tilde{\alpha}=7.315334297\times10^{-3}$
Expected value of the fine structure constant: $\alpha=7.29735588\times10^{-3}$
Expected value of the η constant: $\eta=\frac{\tilde{\mu}_\circ}{\mu_\circ}=0.999901825$

Initial value of the Bohr radius: $\tilde{a}_e = 5.313038333 \times 10^{-11} [m]$

Experimentally received value of the Bohr radius:[2] $a_e = 5.2917706 \times 10^{-11} [m]$

Reduced mass factor: $(1 - \tilde{m}_e / \tilde{m}_p) = 0.999455756$

yields

$$\left(\frac{a_e^2}{\tilde{a}_e^2} \right) (1 - \tilde{m}_e / \tilde{m}_p)^3 \left(\eta^2 \times \frac{\tilde{\alpha}^4}{\alpha^4} \right) = 0.999991$$

This result justifies the definition stated by Eq. 130. Finally, extracting a square root and rearranging yields

$$(131) \qquad a_e = \tilde{a}_e \left[\frac{1}{(1 - \tilde{m}_e / \tilde{m}_p)^{3/2}} \left(\frac{1}{\eta} \right) \left(\frac{\alpha^2}{\tilde{\alpha}^2} \right) \right]$$

Eq. 131 allows for theoretically calculating the Bohr radius a_e, returning the value that matches the experiments.

The definition of the right side of Eq. 130 defines the left side of Eq. 130 as also equal to **1**:

$$\frac{\tilde{m}_e (1 - \tilde{m}_e / \tilde{m}_p)^2}{m_e v_e^2} \left[\frac{\tilde{v}_e^2}{(1 - \tilde{\alpha}^2 / \beta^2)} \right] \left(\frac{a_e}{\tilde{a}_e} \right) \left(\eta \times \frac{\tilde{\alpha}^2}{\alpha^2} \right) = 1$$

or

$$(132) \qquad \frac{\tilde{m}_e \tilde{v}_e^2}{\tilde{a}_e} \left[\frac{(1 - \tilde{m}_e / \tilde{m}_p)^2}{(1 - \tilde{\alpha}^2 / \beta^2)} \right] \left[\eta \times \frac{\tilde{\alpha}^2}{\alpha^2} \right] = \frac{m_e v_e^2}{a_e}$$

Substituting the expression for the Bohr radius a_e from Eq. 131 into Eq. 132, along with the relations $\tilde{v}_e = c\tilde{\alpha}$ and $v_e = c\alpha$ from Eq. 6, allows us to isolate the electron mass m_e and calculate its value, which matches the experiments:

$$(133) \qquad m_e = \tilde{m}_e \left[\frac{(1 - \tilde{m}_e / \tilde{m}_p)^{1/2}}{(1 - \tilde{\alpha}^2 / \beta^2)} \right] \left(\frac{\tilde{\alpha}^2}{\alpha^2} \right)$$

In order to calculate the final theoretical values of a_e and m_e using Eq. 131 and Eq. 133, the final theoretical values of α and constant η should be theoretically determined first, as will be shown.

2-7. The theoretical expressions for the proton mass m_p and proton radius r_p:

Dividing both sides of the equality in Eq. 132 by (v_e^2), and substituting the relations $\tilde{v}_e = c\tilde{\alpha}$ and $v_e = c\alpha$ from Eq. 6, and rearranging yields

$$(134) \qquad \left(\frac{\tilde{m}_e}{\tilde{a}_e} \right) \left[\frac{(1 - \tilde{m}_e / \tilde{m}_p)^2}{(1 - \tilde{\alpha}^2 / \beta^2)} \right] \left[\eta \times \frac{\tilde{\alpha}^4}{\alpha^4} \right] = \frac{m_e}{a_e}$$

Now, multiplying Eq. 33, as follows,

$$\left(\frac{\tilde{a}_e}{\tilde{r}_p}\right)\frac{4\tilde{\alpha}^2}{3\pi}=1$$

by Eq. 32

$$\left(\frac{\tilde{m}_p}{\tilde{m}_e}\right)\frac{4\tilde{\alpha}\beta\sqrt{1-\beta^2}}{3\pi}=1$$

equalizing it to 1, and transposing and rearranging, yields

$$(135)\qquad \frac{\tilde{m}_e}{\tilde{a}_e}=\left(\frac{\tilde{m}_p}{\tilde{r}_p}\right)\frac{16\tilde{\alpha}^3\beta\sqrt{1-\beta^2}}{9\pi^2}$$

Substituting Eq. 135 into the left side of Eq. 134 yields

$$(136)\qquad \left(\frac{\tilde{m}_p}{\tilde{r}_p}\right)\frac{16\ \tilde{\alpha}^3\beta\sqrt{1-\beta^2}}{9\pi^2}\left[\frac{(1-\tilde{m}_e/\tilde{m}_p)^2}{(1-\tilde{\alpha}^2/\beta^2)}\right]\left[\eta\times\frac{\tilde{\alpha}^4}{\alpha^4}\right]=\frac{m_e}{a_e}$$

Now, set Eq. 17a and Eq. 17b equal to each other, as follows

$m_e v_e 2\pi\ a_e = m_p v_p 2\pi\ r_p\sqrt{1-\beta^2}$

Substituting the relations $v_e = c\alpha$ from Eq. 6 and $v_p = c\beta$ from Eq. 7 appropriately on each side, reducing 2π, and rearranging yields

$$(137)\qquad m_e = m_p\left(\frac{\beta}{\alpha}\right)\left(\frac{r_p}{a_e}\right)\sqrt{1-\beta^2}$$

Multiplying both sides of Eq. 137 by r_p/a_e yields

$$m_e\left(\frac{r_p}{a_e}\right)=m_p\left(\frac{\beta}{\alpha}\right)\left(\frac{r_p}{a_e}\right)\sqrt{1-\beta^2}\left(\frac{r_p}{a_e}\right)$$

Or, after transposing and rearranging,

$$(138)\qquad \frac{m_e}{a_e}=\left(\frac{m_p}{r_p}\right)\left(\frac{\beta}{\alpha}\right)\left(\frac{r_p^2}{a_e^2}\right)\sqrt{1-\beta^2}$$

Substituting the ratio m_e/a_e from Eq. 138 into the right side of Eq. 136 yields

$$\left(\frac{\tilde{m}_p}{\tilde{r}_p}\right)\frac{16\tilde{\alpha}^3\beta\times\sqrt{1-\beta^2}}{9\pi^2}\left[\frac{(1-\tilde{m}_e/\tilde{m}_p)^2}{(1-\tilde{\alpha}^2/\beta^2)}\right]\left[\eta\times\frac{\tilde{\alpha}^4}{\alpha^4}\right]=\left(\frac{m_p}{r_p}\right)\left(\frac{\beta}{\alpha}\right)\left(\frac{r_p^2}{a_e^2}\right)\sqrt{1-\beta^2}$$

Or, after transposing and rearranging,

$$(139)\quad \left(\frac{\tilde{m}_p}{\tilde{r}_p}\right)\frac{1}{(1-\tilde{\alpha}^2/\beta^2)}\left(\frac{r_p}{m_p}\right)\left[\eta\times\frac{\tilde{\alpha}^4}{\alpha^4}\right]=\left(\frac{r_p^2}{a_e^2}\right)\frac{1}{\alpha}\left[\frac{1}{(1-\tilde{m}_e/\tilde{m}_p)^2}\right]\left(\frac{9\pi^2}{16\tilde{\alpha}^3}\right)$$

Multiplying the right side of Eq. 139 by $\frac{\tilde{\alpha}}{\tilde{\alpha}}$ yields

$$(140)\quad \left(\frac{\tilde{m}_p}{\tilde{r}_p}\right)\frac{1}{(1-\tilde{\alpha}^2/\beta^2)}\left(\frac{r_p}{m_p}\right)\left(\eta\times\frac{\tilde{\alpha}^4}{\alpha^4}\right)=\left(\frac{r_p^2}{a_e^2}\right)\frac{\tilde{\alpha}}{\alpha}\left[\frac{1}{(1-\tilde{m}_e/\tilde{m}_p)^2}\right]\left(\frac{9\pi^2}{16\tilde{\alpha}^4}\right)$$

Substituting Eq. 33, raised to a power of **2,**

$$\frac{9\pi^2}{16\tilde{\alpha}^4}=\frac{\tilde{a}_e^2}{\tilde{r}_p^2}$$

into the right side of Eq. 140 yields

$$(141)\quad \left(\frac{\tilde{m}_p}{\tilde{r}_p}\right)\frac{1}{(1-\tilde{\alpha}^2/\beta^2)}\left(\frac{r_p}{m_p}\right)\left(\eta\times\frac{\tilde{\alpha}^4}{\alpha^4}\right)=\left(\frac{r_p^2}{a_e^2}\right)\frac{\tilde{\alpha}}{\alpha}\left[\frac{1}{(1-\tilde{m}_e/\tilde{m}_p)^2}\right]\left(\frac{\tilde{a}_e^2}{\tilde{r}_p^2}\right)$$

Defining the right side of Eq. 141 as equal to **1**

$$\left(\frac{r_p^2}{a_e^2}\right)\frac{\tilde{\alpha}}{\alpha}\left[\frac{1}{(1-\tilde{m}_e/\tilde{m}_p)^2}\right]\left(\frac{\tilde{a}_e^2}{\tilde{r}_p^2}\right)=1$$

allows us to isolate an expression that contains the ratio r_p/a_e, from which it is possible to calculate the proton radius r_p.

Substituting the following data values into the right side of Eq. 141

Initial value of the fine structure constant: $\tilde{\alpha}=7.315334297\times10^{-3}$

Expected value of the fine structure constant: $\alpha=7.29735588\times10^{-3}$

Initial value of the Bohr radius: $\tilde{a}_e=5.313038333\times10^{-11}[m]$

Experimentally received value of the Bohr radius:[2] $a_e=5.2917706\times10^{-11}[m]$

Initial value of the proton radius: $\tilde{r}_p=1.206702376\times10^{-15}[m]$

Reduced mass factor: $(1-\tilde{m}_e/\tilde{m}_p)=0.999455756$

Estimated experimentaly received value of the proton radius: $r_p=R_\circ\approx1.2\times10^{-15}[m]$

yields

$$\left(\frac{r_p^2}{a_e^2}\right)\frac{\tilde{\alpha}}{\alpha}\left[\frac{1}{(1-\tilde{m}_e/\tilde{m}_p)^2}\right]\left(\frac{\tilde{a}_e^2}{\tilde{r}_p^2}\right)=1.000431$$

This result justifies the definition stated by Eq. 141. Finally, extracting a square root and rearranging yields

$$(142) \qquad r_p = \tilde{r}_p \left[\frac{a_e(1 - \tilde{m}_e / \tilde{m}_p)}{\tilde{a}_e} \right] \sqrt{\frac{\alpha}{\tilde{\alpha}}}$$

This allows for calculating the *non-relativistic proton radius* r_p. In order to produce the value of the proton radius that matches the experiments, it should be multiplied by $\sqrt{1-\beta^2}$, as shown in Eq. 4:

$$r_p' = r_p \sqrt{1 - v_p^2 / c^2}$$

The definition of the right side of Eq. 141 defines the left side of Eq. 141 as equal to **1,** as well:

$$(143) \qquad \left(\frac{\tilde{m}_p}{\tilde{r}_p} \right) \frac{1}{(1 - \tilde{\alpha}^2 / \beta^2)} \left(\frac{r_p}{m_p} \right) \left(\eta \times \frac{\tilde{\alpha}^4}{\alpha^4} \right) = 1$$

Setting Eq. 17a equal to Eq. 17b

$$m_p v_p 2\pi\, r_p \sqrt{1 - \beta^2} = m_e v_e 2\pi\, a_e$$

and raising both sides to a power of **2**, and transposing and rearranging, yields

$$\frac{r_p^2}{a_e^2} = \left(\frac{m_e^2}{m_p^2} \right) \left(\frac{\alpha^2}{\beta^2} \right) \frac{1}{(1 - \beta^2)}$$

Then, substituted into Eq. 139, yields

$$\left(\frac{\tilde{m}_p}{\tilde{r}_p} \right) \frac{1}{(1 - \tilde{\alpha}^2 / \beta^2)} \left(\frac{r_p}{m_p} \right) \left[\eta \times \frac{\tilde{\alpha}^4}{\alpha^4} \right] = \left(\frac{m_e^2}{m_p^2} \right) \left(\frac{\alpha^2}{\beta^2} \right) \frac{1}{(1 - \beta^2)} \frac{1}{\alpha} \left[\frac{1}{(1 - \tilde{m}_e / \tilde{m}_p)^2} \right] \left(\frac{9\pi^2}{16\tilde{\alpha}^3} \right)$$

or

$$(144) \qquad \left(\frac{\tilde{m}_p}{\tilde{r}_p} \right) \frac{1}{(1 - \tilde{\alpha}^2 / \beta^2)} \left(\frac{r_p}{m_p} \right) \left[\eta \times \frac{\tilde{\alpha}^4}{\alpha^4} \right] = \left(\frac{m_e^2}{m_p^2} \right) \frac{\alpha}{\tilde{\alpha}} \left[\frac{1}{(1 - \tilde{m}_e / \tilde{m}_p)^2} \right] \left[\frac{9\pi^2}{16\tilde{\alpha}^2 \beta^2 (1 - \beta^2)} \right]$$

The last term in the brackets on the right side of Eq. 144 is the inverse of Eq. 32 (raised to a power of **2**), as here

$$\frac{4\tilde{\alpha}\beta\sqrt{1 - \beta^2}}{3\pi} = \left(\frac{\tilde{m}_e}{\tilde{m}_p} \right)$$

Substituting the inverse of Eq. 32, raised to a power of **2,** into Eq. 144, while reducing as needed and rearranging the equation, yields

$$(145) \qquad \left(\frac{\tilde{m}_p}{\tilde{r}_p} \right) \frac{1}{(1 - \tilde{\alpha}^2 / \beta^2)} \left(\frac{r_p}{m_p} \right) \left[\eta \times \frac{\tilde{\alpha}^4}{\alpha^4} \right] = \left(\frac{m_e^2}{m_p^2} \right) \frac{\alpha}{\tilde{\alpha}} \left[\frac{1}{(1 - \tilde{m}_e / \tilde{m}_p)^2} \right] \left(\frac{\tilde{m}_p^2}{\tilde{m}_e^2} \right)$$

The left side of Eq. 145 is identical to Eq. 143**,** as shown here

$$\left(\frac{\tilde{m}_p}{\tilde{r}_p}\right)\frac{1}{(1-\tilde{\alpha}^2/\beta^2)}\left(\frac{r_p}{m_p}\right)\left[\eta\times\frac{\tilde{\alpha}^4}{\alpha^4}\right]=1$$

Thus, setting the right side of Eq. 145 equal to **1,**

(146) $$\left(\frac{m_e^2}{m_p^2}\right)\frac{\alpha}{\tilde{\alpha}}\left[\frac{1}{(1-\tilde{m}_e/\tilde{m}_p)^2}\right]\left(\frac{\tilde{m}_p^2}{\tilde{m}_e^2}\right)=1$$

isolating the proton mass m_p in Eq. 146, and extracting the square root, yields

(147) $$m_p=\tilde{m}_p\left[\frac{m_e}{\tilde{m}_e(1-\tilde{m}_e/\tilde{m}_p)}\right]\sqrt{\frac{\alpha}{\tilde{\alpha}}}$$

Eq. 147 allows for calculating the exact theoretical value of the proton mass m_p, which matches the experiments. The theoretical calculation of the proton radius and mass will be possible only after finding the theoretical values of the Bohr radius and the electron mass, using their theoretical values expressions from the previous development later on.

3
The Calculation of the η Constant

The factor γ of the electron, derived from the electron velocity at the Bohr radius, is described as

(148) $$\gamma_{electron} = \sqrt{1-\tilde{v}_e^2/c^2} = \sqrt{1-\tilde{\alpha}^2}$$

Entering the following data value of the initial fine structure constant

$$\tilde{\alpha} = 7.315334297\times 10^{-3}$$

into Eq. 148 yields

$$\gamma_{electron} = \sqrt{1-\tilde{\alpha}^2} = 0.999973242$$

This allows for presenting the relativistic Bohr radius

(149) $$\tilde{a}_e\sqrt{1-\tilde{v}_e^2/c^2} = \tilde{a}_e\sqrt{1-\tilde{\alpha}^2}$$

Now, going back to Eq. 132

$$\frac{\tilde{m}_e\tilde{v}_e^{\,2}}{\tilde{a}_e}\left[\frac{(1-\tilde{m}_e/\tilde{m}_p)^2}{(1-\tilde{\alpha}^2/\beta^2)}\right]\left[\eta\times\frac{\tilde{\alpha}^2}{\alpha^2}\right] = \frac{m_e v_e^{\,2}}{a_e}$$

and multiplying both sides by $1/\sqrt{1-\tilde{\alpha}^2}$, and dividing both sides by 4π, which was reduced early in the development, yields

(150) $$\frac{\tilde{m}_e\tilde{v}_e^{\,2}}{4\pi\left(\tilde{a}_e\sqrt{1-\tilde{\alpha}^2}\right)}\left[\frac{(1-\tilde{m}_e/\tilde{m}_p)^2}{(1-\tilde{\alpha}^2/\beta^2)}\right]\left[\eta\times\frac{\tilde{\alpha}^2}{\alpha^2}\right] = \frac{m_e v_e^{\,2}}{4\pi\, a_e\sqrt{1-\tilde{\alpha}^2}}$$

Multiplying the right side of Eq. 150 by c_n^2/c_n^2, and substituting the relations $\tilde{v}_e = c\tilde{\alpha}$ and $v_e = c\alpha$ accordingly, yields

(151) $$\frac{\tilde{m}_e c^2\tilde{\alpha}^2}{4\pi\left(\tilde{a}_e\sqrt{1-\tilde{\alpha}^2}\right)}\left[\frac{(1-\tilde{m}_e/\tilde{m}_p)^2}{(1-\tilde{\alpha}^2/\beta^2)}\right]\left[\eta\times\frac{\tilde{\alpha}^2}{\alpha^2}\right] = \frac{m_e c^2\alpha^2}{4\pi\, a_e\sqrt{1-\tilde{\alpha}^2}}\left(\frac{c_n^2}{c_n^2}\right)$$

Transposing and rearranging Eq. 151, without reducing components, yields

(152) $$\frac{\tilde{\alpha}}{4\pi\left(\tilde{a}_e\sqrt{1-\tilde{\alpha}^2}\right)\sqrt{1-\tilde{\alpha}^2/\beta^2}} = \left(\frac{m_e}{\tilde{m}_e}\right)\frac{\alpha}{4\pi\, a_e\sqrt{1-\tilde{\alpha}^2}}\left[\frac{\sqrt{1-\tilde{\alpha}^2/\beta^2}}{(1-\tilde{m}_e/\tilde{m}_p)^2}\right]\frac{1}{\eta}\left(\frac{\alpha^3}{\tilde{\alpha}^3}\right)\left(\frac{c_n^2}{c^2}\right)\left(\frac{c^2}{c_n^2}\right)$$

Now, dividing both sides of Eq. 133 by $\widetilde{m}_e$ yields

$$\frac{m_e}{\widetilde{m}_e}=\left[\frac{(1-\widetilde{m}_e/\widetilde{m}_p)^{1/2}}{(1-\widetilde{\alpha}^2/\beta^2)}\right]\left(\frac{\widetilde{\alpha}^2}{\alpha^2}\right)$$

Substituting it into Eq. 152 and multiplying both sides by $\frac{c_n}{c}$, reducing and rearranging, yields

$$\text{(153)}\qquad \frac{\widetilde{\alpha}}{4\pi\left(\widetilde{a}_e\sqrt{1-\widetilde{\alpha}^2}\right)\sqrt{1-\widetilde{\alpha}^2/\beta^2}}\left(\frac{c_n}{c}\right)$$

$$=\frac{\alpha}{4\pi\ a_e}\left[\left(\frac{c_n^2}{c^2}\right)\left(\frac{\alpha}{\widetilde{\alpha}}\right)\frac{1}{\sqrt{1-\widetilde{m}_e/\widetilde{m}_p}\sqrt{1-\widetilde{\alpha}^2/\beta^2}}\right]\left[\frac{1}{\sqrt{1-\widetilde{\alpha}^2}\,(1-\widetilde{m}_e/\widetilde{m}_p)}\left(\frac{1}{\eta}\right)\left(\frac{c}{c_n}\right)\right]$$

The second term in the brackets on the right side of Eq. 153 is equal to **1**, according to Eq. 127

$$\left[\left(\frac{c_n^2}{c^2}\right)\left(\frac{\alpha}{\widetilde{\alpha}}\right)\frac{1}{\sqrt{1-\widetilde{m}_e/\widetilde{m}_p}\sqrt{1-\widetilde{\alpha}^2/\beta^2}}\right]=1$$

or

$$\frac{c^2}{c_n^2}=\left(\frac{\alpha}{\widetilde{\alpha}}\right)\frac{1}{\sqrt{1-\widetilde{m}_e/\widetilde{m}_p}\sqrt{1-\widetilde{\alpha}^2/\beta^2}}$$

Defining the third term in the brackets on the right side of Eq. 153 as equal to **1** allows for isolating an expression that includes the ratio $\frac{c}{c_n}$, from which it is possible to further isolate the η constant

$$\left[\frac{1}{\sqrt{1-\widetilde{\alpha}^2}\,(1-\widetilde{m}_e/\widetilde{m}_p)}\left(\frac{1}{\eta}\right)\left(\frac{c}{c_n}\right)\right]=1$$

or

$$\text{(154)}\qquad \frac{c}{c_n}=\sqrt{1-\widetilde{\alpha}^2}\left(1-\widetilde{m}_e/\widetilde{m}_p\right)\eta$$

Having Eq. 109 presented in the following form,

$$\frac{e}{\widetilde{e}}=\frac{\left(1-\widetilde{m}_e/\widetilde{m}_p\right)^3}{\eta^3}$$

substituting it into Eq. 126

$$\frac{c^2}{c_n^2}=\frac{e}{\widetilde{e}}$$

while extracting the square root, yields

(155) $$\frac{c}{c_n} = \left[\frac{\left(1 - \widetilde{m}_e / \widetilde{m}_p\right)^3}{\eta^3}\right]^{1/2}$$

Equating Eq. 155 and Eq. 154

$$\left[\frac{\left(1 - \widetilde{m}_e / \widetilde{m}_p\right)^3}{\eta^3}\right]^{1/2} = \sqrt{1 - \widetilde{\alpha}^2}\left(1 - \widetilde{m}_e / \widetilde{m}_p\right) \eta$$

raising both sides of the equality to a power of **2,** and transposing to isolate the constant η, yields

$$\eta^5 = \frac{\left(1 - \widetilde{m}_e / \widetilde{m}_p\right)}{\left(1 - \widetilde{\alpha}^2\right)}$$

or

(156) $$\eta = \left[\frac{\left(1 - \widetilde{m}_e / \widetilde{m}_p\right)}{\left(1 - \widetilde{\alpha}^2\right)}\right]^{1/5}$$

Substituting the following data values in Eq. 156

Initial value of the fine structure constant: $\widetilde{\alpha} = 7.315334297 \times 10^{-3}$
Initial value of the electron mass: $\widetilde{m}_e = 9.051816380 \times 10^{-31} [kg]$
Initial value of the proton mass: $\widetilde{m}_p = 1.663190691 \times 10^{-27} [kg]$

Yields the *theoretical value of the constant* η

$\eta = 0.9999018295$

For comparison, the expected value calculated for the η constant, based on the experimentaly received value [2] of the permeability of vacuum constant $\mu_{\circ}$ and permittivity of vacuum constant $\varepsilon_{\circ}$, is

$\eta = 0.999901825$

This result justifies the definition stated by Eq. 153.

Finally, extracting a square root and rearranging yields

$$\eta = \left[\frac{1}{\sqrt{1 - \widetilde{\alpha}^2}\,(1 - \widetilde{m}_e / \widetilde{m}_p)}\left(\frac{c}{c_n}\right)\right]$$

Substituting the relation $\eta = \dfrac{\varepsilon_{\circ}}{\widetilde{\varepsilon}_{\circ}}$ and rearranging the equation yields an additional expression for calculating the theoretical value of the permittivity of vacuum constant $\varepsilon_{\circ}$, which matches the experiments

(157a) $$\varepsilon_{\circ} = \frac{\widetilde{\varepsilon}_{\circ}}{\sqrt{1 - \widetilde{\alpha}^2}\left(1 - \widetilde{m}_e / \widetilde{m}_p\right)}\left(\frac{c}{c_n}\right)$$

Based on the identity $\varepsilon_{\circ} = 1/\mu_{\circ}c^2$ from Maxwell relations, an additional equation for permeability of vacuum constant $\mu_{\circ}$, which matches the experiments, is obtained

$$\mu_{\circ} = \tilde{\mu}_{\circ}\sqrt{1-\tilde{\alpha}^2}\left(1-\tilde{m}_e/\tilde{m}_p\right)\left(\frac{c_n}{c}\right) \qquad (157b)$$

The theoretical value of the η constant is the last piece of information needed in order to calculate the theoretical values of the different constants, which match their experimentally received values based on all of the expressions developed so far.

4
Calculating the Theoretical Values of the Fundamental Constants

This chapter presents the calculation of the different fundamental constants values, based on the theoretical expressions developed so far, which yields values that match their experimentally received values.

(1) The theoretical value of the permeability of vacuum constant $\mu_\circ$:

Substituting the theoretical value of the η constant calculated from Eq. 156

$$\eta = 0.9999018295$$

and the initial value of the permeability of vacuum constant calculated from Eq. 80

$$\tilde{\mu}_\circ = 1.256513691 \times 10^{-6} \left[N\ A^{-2}\right]$$

into the relation $\eta = \dfrac{\tilde{\mu}_\circ}{\mu_\circ}$, yields the theoretical value of the permeability of vacuum constant $\mu_\circ$

$$\mu_\circ = 1.256637056 \times 10^{-6} \left[H\ m^{-1}\right]$$

For comparison, the experimentally received value [2] of the permeability of vacuum constant $\mu_\circ$ is

$$\mu_\circ = 1.25663706 \times 10^{-6} \left[H\ m^{-1}\right]$$

(2) The theoretical value of the permittivity of vacuum constant $\varepsilon_\circ$:

Substituting the theoretical value of the speed of light in vacuum c calculated from Eq. 96

$$c = 2.99792344 \times 10^{8} \left[m/s\right]$$

and the theoretical value of the permeability of vacuum constant $\mu_\circ$, into $\varepsilon_\circ = 1/\mu_\circ c^2$ from Maxwell relations, yields the theoretical value of the permittivity of vacuum constant $\varepsilon_\circ$

$$\varepsilon_\circ = 8.854194546 \times 10^{-12} \left[F\ m^{-1}\right]$$

For comparison, the experimentally received value [2] of the permittivity of vacuum constant $\varepsilon_\circ$ is

$$\varepsilon_\circ = 8.854187818 \times 10^{-12} \left[F\ m^{-1}\right]$$

(3) The theoretical value of the elementary charge e:

Substituting the theoretical value of the η constant
$\eta = 0.9999018295$
the initial value of the elementary charge $\tilde{e}$
$\tilde{e} = 1.60433419 \times 10^{-19} [C]$
and the theoretical reduced mass factor value
$(1 - \tilde{m}_e / \tilde{m}_p) = 0.999455756$
into Eq. 109, here

$$e = \tilde{e}\left(\frac{(1 - \tilde{m}_e / \tilde{m}_p)^3}{\eta^3} \right)$$

Yields the theoretical value of the *elementary charge* e

$$e = 1.602188 \times 10^{-19} [C]$$

For comparison, the experimentally received value [2] of the elementary charge e is

$$e = 1.6021892 \times 10^{-19} [C]$$

(4) The theoretical value of the *fine structure constant* α:

Using Eq. 117, here

$$\alpha = \tilde{\alpha} \frac{\sqrt{1 - \tilde{\alpha}^2 / \beta^2}\,(1 - \tilde{m}_e / \tilde{m}_p)^{7/2}}{\eta^3}$$

Substituting the following data values into Eq. 117
Initial value of the fine structure constant: $\tilde{\alpha} = 7.315334297 \times 10^{-3}$
Reduced mass factor: $(1 - \tilde{m}_e / \tilde{m}_p) = 0.999455756$
Theoretical value of the η constant: $\eta = 0.9999018295$
Effective value of γ factor: $\sqrt{1 - \tilde{\alpha}^2 / \beta^2} = \frac{1}{192\pi^4 \tilde{\alpha}^2} = 0.999150536$
Yields the theoretical value of α

$$\alpha = \frac{1}{137.03593859}$$

For comparison, the experimentally received value [3] of the fine structure constant is

$$\alpha = \frac{1}{137.0360}$$

It also possible to calculate the ratio $\frac{\alpha^2}{\tilde{\alpha}^2}$, the relation between the theoretical value of the fine structure constant α and its initial value $\tilde{\alpha}$,

$$\frac{\alpha^2}{\tilde{\alpha}^2} = 0.9950907466$$

(5) The theoretical value of the *Bohr radius* a_e:
Using Eq. 131, here

$$a_e = \frac{\tilde{a}_e}{\eta \times (1 - \tilde{m}_e / \tilde{m}_p)^{3/2}} \times \frac{\alpha^2}{\tilde{\alpha}^2}$$

Substituting the following data values into Eq. 131
Initial value of the Bohr radius: $\tilde{a}_e = 5.313038333 \times 10^{-11} [m]$
Theoretical value of the reduced mass factor: $(1 - \tilde{m}_e / \tilde{m}_p) = 0.999455756$
Theoretical value of the η constant: $\eta = 0.9999018295$
Ratio of $\frac{\alpha^2}{\tilde{\alpha}^2} = 0.9950907466$
Yields the theoretical value of the *Bohr radius* a_e

$$a_e = 5.291793809 \times 10^{-11} [m]$$

For comparison, the experimentally received value [2] of the Bohr radius is

$$a_e = 5.2917706 \times 10^{-11} [m]$$

(6) The theoretical value of the *electron mass* m_e:
Using Eq. 133, here

$$m_e = \tilde{m}_e \left[\frac{(1 - \tilde{m}_e / \tilde{m}_p)^{1/2}}{(1 - \tilde{\alpha}^2 / \beta^2)} \right] \left(\frac{\tilde{\alpha}^2}{\alpha^2} \right)$$

Substituting the following data values into Eq. 133
Initial value of the electron mass: $\tilde{m}_e = 9.051816380 \times 10^{-31} [kg]$
Initial value of the fine structure constant: $\tilde{\alpha} = 7.315334297 \times 10^{-3}$
Theoretical value of the fine structure constant: $\alpha = \frac{1}{137.03593859}$
Effective value of γ factor: $\sqrt{1 - \tilde{\alpha}^2 / \beta^2} = \frac{1}{192\pi^4 \tilde{\alpha}^2} = 0.999150536$
Reduced mass factor: $(1 - \tilde{m}_e / \tilde{m}_p) = 0.999455756$

Yields the theoretical value of the *electron mass* m_e

$m_e = 9.109467338 \times 10^{-31} [kg]$

For comparison, the experimentally received value [2] of the electron mass is

$m_e = 9.109534 \times 10^{-31} [kg]$

(7) The *theoretical value of the speed of light in the atomic domain* c_n:

Using Eq. 122, here

$$c_n^2 = \frac{\beta}{\mu_\circ^3 (1 - \tilde{m}_e / \tilde{m}_p)^3 \sqrt{1 - \tilde{\alpha}^2 / \beta^2}}$$

Substituting the following data values into Eq. 122
Key constant: $\beta = 0.178145016$
Theoretical permeability of vacuum constant: $\mu_\circ = 1.256637056 \times 10^{-6} [H\ m^{-1}]$

Effective value of γ factor: $\sqrt{1 - \tilde{\alpha}^2 / \beta^2} = \dfrac{1}{192\pi^4 \tilde{\alpha}^2} = 0.999150536$

Reduced mass factor: $(1 - \tilde{m}_e / \tilde{m}_p) = 0.999455756$
Yields, the value of the *speed of light in the atomic domain* c_n
$c_n = 2.999930699 \times 10^8 [m\ s^{-1}] = 299993.0699 [km\ s^{-1}]$

(8) The *theoretical value of the Rydberg constant* R_∞:

$$R_\infty = \frac{\alpha}{4\pi a_e} = \frac{\tilde{\alpha}}{4\pi \left(\tilde{a}_e \sqrt{1 - \tilde{\alpha}^2}\right) \sqrt{1 - \tilde{\alpha}^2 / \beta^2}} \left(\frac{c_n}{c}\right)$$

or

$$R_\infty = \frac{\tilde{\alpha}}{4\pi \tilde{a}_e \sqrt{1 - \tilde{\alpha}^2} \sqrt{1 - \tilde{\alpha}^2 / \beta^2}} \left(\frac{c_n}{c}\right)$$

The last equation is the *Rydberg constant* R_∞, derived from a more general expression.[8] This enables the right side of the equality to become an additional expression for finding the *Rydberg constant.*

Substituting the following data values into the last equation

Initial value of the fine structure constant: $\tilde{\alpha} = 7.315334297 \times 10^{-3}$
Theoretical value of speed of light in the atomic domain: $c_n = 2.999930699 \times 10^8 [m\ s^{-1}]$
Theoretical speed of light in vacuum: $c = 2.99792344 \times 10^8 [m/s]$
Initial value of the Bohr radius: $\tilde{a}_e = 5.313038333 \times 10^{-11} [m]$
Key constant: $\beta = 0.178145016$
Yields the theoretical value of *Rydberg constant* R_∞

$$R_\infty = 1.097369141 \times 10^{7} \left[m^{-1}\right]$$

For comparison, the experimentally received value [2] of Rydberg constant is

$$R_\infty = 1.097373177 \times 10^{7} \left[m^{-1}\right]$$

(9) The theoretical value of the *non-relativistic proton radius* r_p :

Using Eq. 142

$$r_p = \tilde{r}_p a_e \left[\frac{(1 - \tilde{m}_e / \tilde{m}_p)}{\tilde{a}_e}\right] \sqrt{\frac{\alpha}{\tilde{\alpha}}}$$

Substituting the following data values into Eq. 142
Theoretical value of the Bohr radius: $a_e = 5.291793809 \times 10^{-11} [m]$
Initial value of the Bohr radius: $\tilde{a}_e = 5.313038333 \times 10^{-11} [m]$
Reduced mass factor: $(1 - \tilde{m}_e / \tilde{m}_p) = 0.99945575\ 6$
Theoretical fine structure constant: $\alpha = \dfrac{1}{137.03593859}$
Initial value of the fine structure constant: $\tilde{\alpha} = 7.315334297 \times 10^{-3}$
Initial proton radius: $\tilde{r}_p = 1.206702376 \times 10^{-15} [m]$
Yields the theoretical *non-relativistic proton radius*

$$r_p = 1.199746186 \times 10^{-15} [m]$$

(10) The theoretical value of the *relativistic proton radius* r_p'

Using the correction expression $\sqrt{1-\beta^2}$ from Eq. 4 yields the *relativistic proton radius* r_p'

$$r_p' = r_p \sqrt{1-\beta^2} = 1.180555339 \times 10^{-15} [m]$$

(11) The theoretical value of the *proton mass* m_p :

Using Eq. 147, here

$$m_p = \tilde{m}_p \left[\frac{m_e}{\tilde{m}_e (1 - \tilde{m}_e / \tilde{m}_p)}\right] \sqrt{\frac{\alpha}{\tilde{\alpha}}}$$

Substituting the following data values into Eq. 147
Theoretical value of the fine structure constant: $\alpha = \dfrac{1}{137.03593859}$
Initial value of the fine structure constant: $\tilde{\alpha} = 7.315334297 \times 10^{-3}$
Theoretical value of the electron mass: $m_e = 9.109467338 \times 10^{-31} [kg]$
Initial value of the electron mass: $\tilde{m}_e = 9.051816380 \times 10^{-31} [kg]$

Reduced mass factor: $(1-\tilde{m}_e/\tilde{m}_p)=0.999455756$
Initial proton mass: $\tilde{m}_p=1.663190691\times10^{-27}[kg]$
Yields the theoretical value of the *proton mass* m_p

$$m_p=1.672635814\times10^{-27}[kg]$$

For comparison, the experimentally received value [2] of the proton mass is

$$m_p=1.67262311\times10^{-27}[kg]$$

(12) The theoretical value of the *Compton wavelength for the electron* $\lambda_{C,e}$:
Using Eq. 17a, along with the expression $v_e=c\alpha$ from Eq. 6

$$\lambda_{C,e}=\frac{h}{m_e c}=2\pi a_e\alpha$$

Substituting the following data values into the last equation

Theoretical value of the fine structure constant: $\alpha=\dfrac{1}{137.03593859}$

Theoretical value of the Bohr radius: $a_e=5.291793809\times10^{-11}[m]$

Yields the theoretical value of the *Compton wavelength for the electron* $\lambda_{C,e}$

$$\lambda_{C,e}=2.426321259\times10^{-12}[m]$$

For comparison, the experimentally received value [2] is

$$\lambda_{C,e}=2.426310\times10^{-12}[m]$$

(13) The theoretical value of the *Compton wavelength for the proton* $\lambda_{C,p}$:
Using Eq. 17b, along with the expression $v_p=c\beta$ from Eq. 7

$$\lambda_{C,p}=\frac{h}{m_p c}=2\pi\beta\, r_p\sqrt{1-\beta^2}$$

Substituting the following data values into the last equation

Theoretical values of the key constant: $\beta=0.178145016$
Relativistic proton radius: $r_p'=r_p\sqrt{1-\beta^2}=1.1805\ 5533\ 9\times10^{-15}[m]$
Yields the theoretical value of the *Compton wavelength for the proton* $\lambda_{C,p}$

$$\lambda_{C,p}=1.321417015\times10^{-15}[m]$$

For comparison, the experimentally received value [8] is

$$\lambda_{C,p}=1.32141002\times10^{-15}[m]$$

(14) The theoretical value of the *proton mass to electron mass ratio* $\frac{m_p}{m_e}$**:**

Using Eq. 147 after dividing both sides of the equality by m_e

$$\frac{m_p}{m_e} = \frac{\tilde{m}_p}{\tilde{m}_e(1 - \tilde{m}_e / \tilde{m}_p)} \sqrt{\frac{\alpha}{\tilde{\alpha}}}$$

Substituting the following data values into Eq. 147

Theoretical value of the fine structure constant: $\alpha = \frac{1}{137.03593859}$

Initial value of the fine structure constant: $\tilde{\alpha} = 7.315334297 \times 10^{-3}$

Reduced mass factor: $(1 - \tilde{m}_e / \tilde{m}_p) = 0.999455756$

Initial value of the electron mass: $\tilde{m}_e = 9.051816380 \times 10^{-31} [kg]$

Initial value of the proton mass: $\tilde{m}_p = 1.663190691 \times 10^{-27} [kg]$

Yields the theoretical value of the *proton mass to electron mass ratio* $\frac{m_p}{m_e}$

$$\frac{m_p}{m_e} = 1836.151064$$

For comparison, the experimentally received value [2] is

$$\frac{m_p}{m_e} = 1836.15152$$

(15) The theoretical value of the *gravitation constant* G:

Using Eq. 46, with $1/e$ used as dimensionless, according to the definition stated previously that it is acting as a counting tool,

$$G = \frac{\alpha c^2 r_p \sqrt{1 - \beta^2}}{\left(\frac{1}{e}\right)\left(\frac{1}{e}\right) m_p \beta}$$

Substituting the following data values into Eq. 46

Key constant: $\beta = 0.178145016$

Theoretical value of the proton mass: $m_p = 1.672635814 \times 10^{-27} [kg]$

Theoretical value of the speed of light in vacuum: $c = 2.99792344 \times 10^8 [m/s]$

Theoretical value related to the elementary charge acting as a counting tool: $\left(\frac{1}{e}\right) = 1/1.602188 \times 10^{-19}$

Relativistic proton radius: $r'_p = r_p \sqrt{1 - \beta^2} = 1.180555339 \times 10^{-15} [m]$

Yields the theoretical value of the *gravitation constant* G

$$G = 6.670291593 \times 10^{-11} \left[m^3 s^{-2} Kg^{-1}\right]$$

For comparison, the experimentally received value [2] is

$$G = 6.6720 \times 10^{-11} \left[m^3 s^{-2} Kg^{-1}\right]$$

(16) The theoretical value of *Planck mass* m_{pl}:
Using Eq. 41, here

$$m_{pl} = \left(\frac{1}{e} m_p \beta\right)$$

Substituting the following data values into Eq. 41

Theoretical value of the proton mass: $m_p = 1.672635814 \times 10^{-27} [kg]$

Theoretical value related to the elementary charge acting as a counting tool: $\left(\frac{1}{e}\right) = 1/1.602188 \times 10^{-19}$

Key constant: $\beta = 0.178145016$

Yields the theoretical value of the *Planck mass* m_{pl}
$m_{pl} = 1.85978 \times 10^{-9} [kg]$.
Note that the known value is calculated according to the following formula:[8]

$$m_{pl} = \left(\frac{\hbar c}{G}\right)^{1/2}$$

though apparently it should be calculated according to the following formula:

$$m_{pl} = \left(\frac{\hbar c \alpha}{G}\right)^{1/2}$$

Using this last formula for the known Planck mass yields the value for comparison:

$$m_{pl} = 1.8595441 \times 10^{-9} [kg]$$

(17) The theoretical value of the *Planck length* l_{pl}:
Using Eq. 48, here

$$l_{pl} = \frac{\alpha r_p \sqrt{1-\beta^2}}{\left(\frac{1}{e}\right)}$$

Substituting the following data values into Eq. 48

Theoretical value of the fine structure constant: $\alpha = \dfrac{1}{137.03593859}$

Theoretical value related to the elementary charge acting as a counting tool: $\left(\frac{1}{e}\right) = 1/1.602188 \times 10^{-19}$

Relativistic proton radius: $r'_p = r_p\sqrt{1-\beta^2} = 1.180555339 \times 10^{-15}[m]$

Yields the theoretical value of the *Planck length* l_{pl}

$l_{pl} = 1.38027412 \times 10^{-36}[m]$

Note that the known value is calculated according to the following formula:[8]

$$l_{pl} = \left(\frac{\hbar G}{c^3}\right)^{1/2}$$

though apparently it should be calculated according to the following formula:

$$l_{pl} = \left(\frac{\hbar G\alpha}{c^3}\right)^{1/2}$$

Using this last formula for the known Planck length yields the value for comparison

$$l_{pl} = 1.38045138 \times 10^{-36}[m]$$

(18) The theoretical value of the *Planck time* t_{pl}:

Using Eq. 48 divided by the speed of light in vacuum c, although it might be divided by the speed of light in the atomic domain c_n

$$t_{pl} = \frac{\alpha r_p\sqrt{1-\beta^2}}{\left(\frac{1}{e}\right)c}$$

Substituting the following data values into the last equation

Theoretical value of the fine structure constant: $\alpha = \frac{1}{137.03593859}$

Theoretical value related to the elementary charge acting as a counting tool: $\left(\frac{1}{e}\right) = 1/1.602188 \times 10^{-19}$

Relativistic proton radius: $r'_p = r_p\sqrt{1-\beta^2} = 1.180555339 \times 10^{-15}[m]$

Theoretical value of the speed of light in vacuum: $c = 2.99792344 \times 10^{8}[m/s]$

Yields, the theoretical value of the *Planck time* t_{pl}

$$t_{pl} = 4.60267 \times 10^{-45}[s]$$

Note that the known value is calculated according to the following formula:[8]

$$t_{pl} = \left(\frac{\hbar \times G}{c^5}\right)^{1/2}$$

though apparently it should be calculated according to the following formula:

$$t_{pl} = \left(\frac{\hbar \times \alpha \times G}{c^5} \right)^{1/2}$$

Using this last formula for the known Planck time yields the value for comparison

$$t_{pl} = 4.60486141 \times 10^{-45} [s]$$

(19) The theoretical value of the *Planck constant* h:

Using Eq. 17a, here

$h = 2\pi a_e c\alpha m_e$

Substituting the following data values into Eq. 17a

Theoretical value of the electron mass: $m_e = 9.109467338 \times 10^{-31} [kg]$

Theoretical value of the Bohr radius: $a_e = 5.291793809 \times 10^{-11} [m]$

Theoretical value of the fine structure constant: $\alpha = \dfrac{1}{137.03593859}$

Theoretical value of the speed of light in vacuum: $c = 2.99792344 \times 10^{8} [m/s]$

Yields the theoretical value of the *Planck constant h*

$$h = 6.62615878 \times 10^{-34} [J\ s]$$

For comparison, the known experimental value[2] is

$$h = 6.626176 \times 10^{-34} [J\ s]$$

PART 4
Applying the Theory in Other Areas of Physics

1
The Neutron

This chapter is dedicated to developing the neutron theoretical expressions, and to the theoretical calculation of the neutron parameters values, including mass and radius, which match their experimentally received values.

1-1. A preliminary development for the neutron expressions

Starting with Eq. 121 in this form for e^2

$$e^2 = \left(\frac{\alpha^2}{\tilde{\alpha}^2}\right)\frac{\tilde{e}^2}{\left(1-\tilde{m}_e/\tilde{m}_p\right)\left(1-\tilde{\alpha}^2/\beta^2\right)}$$

and substituting Eq. 18, with the initial components marked,

$$\tilde{e}^2_{proton} = \frac{\tilde{m}_p v_p^2 4\pi\,\tilde{r}_p^2(1-\beta^2)}{6}\left(\frac{\tilde{m}_p}{\tilde{m}_e}\right)$$

into Eq. 121 yields

$$(158)\qquad e^2 = \frac{\tilde{m}_p v_p^2 4\pi\,\tilde{r}_p^2(1-\beta^2)}{2\times 3}\left(\frac{\tilde{m}_p}{\tilde{m}_e}\right)\left(\frac{\alpha^2}{\tilde{\alpha}^2}\right)\frac{1}{(1-\tilde{m}_e/\tilde{m}_p)\left(1-\tilde{\alpha}^2/\beta^2\right)}$$

Now, Eq. 18 must have the same form as was presented by Eq. 22

$$e^2_{proton} = \left(m_p v_p^2 4\pi\,\varepsilon_\circ r_p\sqrt{1-\beta^2}\right)\frac{\alpha}{\beta}$$

Dividing Eq. 158 by Eq. 22 above yields

$$(159)\qquad \frac{\tilde{m}_p\tilde{r}_p^2\sqrt{1-\beta^2}\,\tilde{m}_p\alpha^2\beta}{m_p r_p 6\varepsilon_\circ\alpha\tilde{m}_e\tilde{\alpha}^2\left(1-\tilde{m}_e/\tilde{m}_p\right)\left(1-\tilde{\alpha}^2/\beta^2\right)} = 1$$

Substituting Eq. 21 with the initial components here

$\tilde{m}_p = \dfrac{\tilde{\alpha}\,\tilde{m}_e\tilde{a}_e}{\tilde{r}_p\beta\sqrt{1-\beta^2}}$, instead of one of the $\tilde{m}_p$ components in Eq. 159, along with the relation $\tilde{a}_e = 6\tilde{\varepsilon}_\circ$

from Eq. 15 and $\eta = \dfrac{\varepsilon_\circ}{\tilde{\varepsilon}_\circ}$, yields after rearrangement

(160) $$\left(\frac{\tilde{m}_p}{m_p}\right)\left(\frac{\tilde{r}_p}{r_p}\right) = \eta(1-\tilde{\alpha}^2/\beta^2)(1-\tilde{m}_e/\tilde{m}_p)\left(\frac{\tilde{\alpha}}{\alpha}\right)$$

In addition, transposing the terms in Eq. 120

$$e^2 = \frac{\tilde{m}_e v_e^2 4\pi\, \tilde{a}_e^2}{6\left(1-\tilde{\alpha}^2/\beta^2\right)\left(1-\tilde{m}_e/\tilde{m}_p\right)} \equiv m_e v_e^2 4\pi\, a_e \varepsilon_\circ$$

along with substituting the relations $\tilde{a}_e = 6\tilde{\varepsilon}_\circ$ from Eq. 15 in its initial form and $\eta = \frac{\varepsilon_\circ}{\tilde{\varepsilon}_\circ}$ while reducing as needed, yields

(161) $$\left(\frac{\tilde{m}_e}{m_e}\right)\left(\frac{\tilde{a}_e}{a_e}\right) = \eta(1-\tilde{\alpha}^2/\beta^2)(1-\tilde{m}_e/\tilde{m}_p)$$

Dividing Eq. 160 by Eq. 161 yields

(162) $$\left(\frac{m_p}{\tilde{m}_p}\right)\left(\frac{r_p}{\tilde{r}_p}\right) = \left(\frac{m_e}{\tilde{m}_e}\right)\left(\frac{a_e}{\tilde{a}_e}\right)\left(\frac{\alpha}{\tilde{\alpha}}\right)$$

Multiplying Eq. 134 in the following form

$$\left[\frac{(1-\tilde{m}_e/\tilde{m}_p)^2}{(1-\tilde{\alpha}^2/\beta^2)}\right]\left(\frac{a_e}{\tilde{a}_e}\right)\left(\frac{\tilde{m}_e}{m_e}\right)\left(\frac{\tilde{\alpha}^4}{\alpha^4}\right)\eta = 1$$

by the inverse of Eq. 143

$$\left(\frac{\tilde{r}_p}{\tilde{m}_p}\right)\left(1-\tilde{\alpha}^2/\beta^2\right)\left(\frac{m_p}{r_p}\right)\left(\frac{1}{\eta}\right)\left(\frac{\alpha^4}{\tilde{\alpha}^4}\right) = 1$$

and reducing as needed, yields

(163) $$\left(\frac{m_e}{a_e}\right)\left[\frac{1}{\tilde{m}_e(1-\tilde{m}_e/\tilde{m}_p)}\right]\left[\frac{\tilde{a}_e}{(1-\tilde{m}_e/\tilde{m}_p)}\right] = \left(\frac{\tilde{r}_p}{\tilde{m}_p}\right)\left(\frac{m_p}{r_p}\right)$$

The denominator of the second expression in brackets on the left side of Eq. 163 is the definition of the electron reduced mass according to Eq. 98. The third expression in brackets on the left side is the definition of the Bohr radius under the influence of the reduced mass factor according to Eq. 99. The results of Eq. 162 and Eq. 163 will be used next.

1-2. The theoretical calculation of the neutron mass m_n and neutron radius r_n:

The relationship between the experimentally received value of the neutron mass m_n and the experimentally received value of the electron mass m_e is based on the similarity to the relationship between the experimentally received value of the proton mass m_p and the experimentally received value of the electron mass m_e.

The first similarity is based on the form of Eq. 162, shown here in a symmetrical form with the neutron parameters m_n and r_n

$$(164) \qquad \left(\frac{m_n}{\tilde{m}_p}\right)\left(\frac{r_n}{\tilde{r}_p}\right)=\left(\frac{m_e}{\tilde{m}_e}\right)\left(\frac{a_e}{\tilde{a}_e}\right)\left(\frac{\alpha}{\tilde{\alpha}}\right)$$

The second similarity is based on Eq. 163, shown here with the neutron parameters m_n and r_n

$$(165) \qquad \left(\frac{m_e}{a_e}\right)\left(\frac{\tilde{a}_e(1-\tilde{m}_e/\tilde{m}_p)^2}{\tilde{m}_e}\right)\left(\frac{\tilde{\alpha}^2}{\alpha^2}\right)=\left(\frac{\tilde{r}_p}{\tilde{m}_p}\right)\left(\frac{m_n}{r_n}\right)$$

Note that in order to receive the correct expression there is a need to add the $\left(\frac{\tilde{\alpha}^2}{\alpha^2}\right)$ component at the left side of Eq. 165**,** and the reduced mass component $(1-\tilde{m}_e/\tilde{m}_p)^2$ moves from the denominator to the numerator.

Now, isolating r_n from Eq. 165 yields

$$(166) \qquad r_n=\left(\frac{m_n}{m_e}\right)\left(\frac{\tilde{r}_p}{\tilde{m}_p}\right)\left(\frac{\tilde{m}_e}{\tilde{a}_e}\right)\frac{a_e}{(1-\tilde{m}_e/\tilde{m}_p)^2}\left(\frac{\alpha^2}{\tilde{\alpha}^2}\right)$$

Substituting Eq. 166 into Eq. 164 while reducing as needed and extracting the square root yields

$$(167) \qquad \frac{m_n}{m_e}=\frac{\tilde{m}_p(1-\tilde{m}_e/\tilde{m}_p)}{\tilde{m}_e}\sqrt{\frac{\tilde{\alpha}}{\alpha}}$$

Substituting the following data values into Eq. 167

Initial value of the fine structure constant: $\tilde{\alpha}=7.315334297\times10^{-3}$

Initial value of the proton mass: $\tilde{m}_p=1.663190691\times10^{-27}[kg]$

Initial value of the electron mass: $\tilde{m}_e=9.051816380\times10^{-31}[kg]$

Theoretical value of the fine structure constant: $\alpha=\dfrac{1}{137.03593859}$

Yields the theoretical value of the *neutron mass to electron mass ratio*

$$\frac{m_n}{m_e}=1838.682747$$

For comparison, the experimentally received value [2] for this ratio is

$$\frac{m_n}{m_e}=1838.683662$$

Eq. 167 allows for isolating and calculating the theoretical value of the neutron mass m_n, which matches the experimentally received value, as follows:

$$(168) \qquad m_n = m_e \frac{\tilde{m}_p(1-\tilde{m}_e/\tilde{m}_p)}{\tilde{m}_e}\sqrt{\frac{\tilde{\alpha}}{\alpha}}$$

Substituting the following data values into Eq. 168
Initial value of the proton mass: $\tilde{m}_p = 1.663190691\times10^{-27}[kg]$
Initial value of the electron mass: $\tilde{m}_e = 9.051816380\times10^{-31}[kg]$
Initial value of the fine structure constant: $\tilde{\alpha} = 7.315334297\times10^{-3}$
Theoretical value of the fine structure constant: $\alpha = \frac{1}{137.03593859}$
Theoretical value of the electron mass: $m_e = 9.109467338\times10^{-31}[kg]$

Yields the theoretical value of the *neutron mass* m_n, which matches the experimentally received value

$$m_n = 1.67493205\times10^{-27}[kg]$$

For comparison, the experimentally received value [2] for the neutron mass is

$$m_n = 1.6749543\times10^{-27}[kg]$$

Substituting Eq. 168 into the expression for the neutron radius in Eq. 166 allows us to isolate and calculate the theoretical *non-relativistic neutron radius* r_n

$$(169) \qquad r_n = a_e\left[\frac{\tilde{r}_p}{\tilde{a}_e(1-\tilde{m}_e/\tilde{m}_p)}\right]\left(\frac{\alpha}{\tilde{\alpha}}\right)^{3/2}$$

Substituting the following data values into Eq. 169
Initial value of the Bohr radius: $\tilde{a}_e = 5.313038333\times10^{-11}[m]$
Theoretical value of the Bohr radius: $a_e = 5.291793809\times10^{-11}[m]$
Initial value of the proton radius: $\tilde{r}_p = 1.206702376\times10^{-15}[m]$
Initial value of the proton mass: $\tilde{m}_p = 1.663190691\times10^{-27}[kg]$
Initial value of the electron mass: $\tilde{m}_e = 9.051816380\times10^{-31}[kg]$
Initial value of the fine structure constant: $\tilde{\alpha} = 7.315334297\times10^{-3}$
Theoretical value of the fine structure constant: $\alpha = \frac{1}{137.03593859}$
Yields the theoretical value of the *non-relativistic neutron radius* r_n

$$r_n = 1.19810139\times10^{-15}[m]$$

Multiplying the non-relativistic neutron radius by the relativistic factor $\sqrt{1-\beta^2}$ from Eq. 4 yields the theoretical *relativistic neutron radius value* r_n'

$$r'_n = r_n \times \sqrt{1-\beta^2} = 1.17893686 \times 10^{-15}[m]$$

The relativistic correction is introduced into the calculation because the neutron has the same spinning velocity as the proton!

Now, using an identical equation to Eq. 17, with the neutron components, while placing the relation $v_p = c\beta$ from Eq. 7, yields the *Compton wavelength of the neutron* $\lambda_{C,n}$.

$$(170) \qquad \lambda_{C,n} = \frac{h}{m_n c} = 2\pi\, r_n \beta \sqrt{1-\beta^2}$$

Eq. 170 allows for calculating the theoretical value for the *Compton wavelength of the neutron* $\lambda_{C,n}$
Substituting the following data values into Eq. 170
Theoretical relativistic value of the neutron radius: $r'_n = 1.17893686 \times 10^{-15}[m]$
Key constant: $\beta = 0.178145016$
Yields the theoretical value for the *Compton wavelength of the neutron*

$$\lambda_{C,n} = 1.319605422 \times 10^{-15}[m]$$

For comparison, the experimentally received value [8] is

$$\lambda_{C,n} = 1.31959110 \times 10^{-15}[m]$$

Eq. 162 is equals to Eq. 164 according to the following:

$$(171) \qquad \left(\frac{m_n}{\tilde{m}_p}\right)\left(\frac{r_n}{\tilde{r}_p}\right) = \left(\frac{m_p}{\tilde{m}_p}\right)\left(\frac{r_p}{\tilde{r}_p}\right) = \left(\frac{m_e}{\tilde{m}_e}\right)\left(\frac{a_e}{\tilde{a}_e}\right)\left(\frac{\alpha}{\tilde{\alpha}}\right)$$

This leads to the relation

$$m_n r_n = m_p r_p$$

or

$$r_n = \frac{m_p r_p}{m_n}$$

Dividing Eq. 168 by Eq. 169 yields

$$(172) \qquad \frac{m_n}{r_n} = \left(\frac{m_e}{\tilde{m}_e}\right)\left(\frac{\tilde{a}_e}{a_e}\right)\left(\frac{\tilde{m}_p}{\tilde{r}_p}\right)\left(\frac{\tilde{\alpha}^2}{\alpha^2}\right)\left(1-\tilde{m}_e/\tilde{m}_p\right)^2$$

Substituting the ratio $\left(\frac{m_e}{\tilde{m}_e}\right)$ from Eq. 133 here

$$\frac{m_e}{\tilde{m}_e} = \left[\frac{(1-\tilde{m}_e/\tilde{m}_p)^{1/2}}{(1-\tilde{\alpha}^2/\beta^2)}\right]\left(\frac{\tilde{\alpha}^2}{\alpha^2}\right)$$

and the ratio $\left(\frac{\tilde{a}_e}{a_e}\right)$ from Eq. 131, here

$$\frac{\tilde{a}_e}{a_e} = \left(\frac{\tilde{\alpha}^2}{\alpha^2}\right)\left(1-\tilde{m}_e/\tilde{m}_p\right)^{3/2}\eta$$

and the ratio $\left(\frac{\tilde{m}_p}{\tilde{r}_p}\right)$ from Eq. 143, as follows

$$\left(\frac{\tilde{m}_p}{\tilde{r}_p}\right) = \frac{1}{\eta}(1-\tilde{\alpha}^2/\beta^2)\left(\frac{m_p}{r_p}\right)\left(\frac{\alpha^4}{\tilde{\alpha}^4}\right)$$

into the right side of Eq. 172, yields, after reducing and rearranging,

(173) $$\frac{m_n}{r_n} = \left(\frac{m_p}{r_p}\right)\left(1-\tilde{m}_e/\tilde{m}_p\right)^4\left(\frac{\tilde{\alpha}^2}{\alpha^2}\right)$$

Substituting the relation $r_n = \frac{m_p r_p}{m_n}$ from Eq. 171 into Eq. 173 yields

(174) $$m_n^2 = m_p^2\left(1-\tilde{m}_e/\tilde{m}_p\right)^4\left(\frac{\tilde{\alpha}^2}{\alpha^2}\right)$$

After extracting the square root from Eq. 174, an additional expression for the *ratio between neutron mass and the proton mass* is produced

$$\frac{m_n}{m_p} = \left(1-\tilde{m}_e/\tilde{m}_p\right)^2\frac{\tilde{\alpha}}{\alpha}$$

Or, isolating the *neutron mass* m_n

(175) $$m_n = m_p\left(1-\tilde{m}_e/\tilde{m}_p\right)^2\frac{\tilde{\alpha}}{\alpha}$$

and substituting Eq. 175 into Eq. 171 yields an additional expression for the ratio between *neutron radius* r_n and the *proton radius* r_p

$$\frac{r_n}{r_p} = \left[\frac{1}{\left(1-\tilde{m}_e/\tilde{m}_p\right)^2}\right]\frac{\alpha}{\tilde{\alpha}}$$

Or, isolating the neutron radius r_n,

(176) $$r_n = r_p \left[\frac{1}{(1 - \tilde{m}_e / \tilde{m}_p)^2} \right] \frac{\alpha}{\tilde{\alpha}}$$

Now, multiplying Eq. 167, here

$$\frac{m_n}{m_e} = \frac{\tilde{m}_p (1 - \tilde{m}_e / \tilde{m}_p)}{\tilde{m}_e} \sqrt{\frac{\tilde{\alpha}}{\alpha}}$$

by Eq. 147, here

$$\frac{m_p}{m_e} = \frac{1}{(1 - \tilde{m}_e / \tilde{m}_p)} \left(\frac{\tilde{m}_p}{\tilde{m}_e} \right) \sqrt{\frac{\alpha}{\tilde{\alpha}}}$$

yields, after reducing and extracting the square root, an interesting equation

(177) $$\sqrt{\left(\frac{m_p}{m_e} \right) \left(\frac{m_n}{m_e} \right)} = \frac{\tilde{m}_p}{\tilde{m}_e}$$

The ratio between the initial proton mass $\tilde{m}_p$ and the initial electron mass $\tilde{m}_e$ is equal to the *geometric square root* of the multiplication between the final theoretical values, the proton to electron mass ratio and neutron to electron mass ratio, namely the values that match the experiments!

2

The Josephson Frequency-Voltage Parameter and Hall Effect

The development starts with Eq. 61, here

$$\tilde{e} = \frac{48\tilde{\alpha}^2}{c^2\beta}$$

Substituting this equation into Eq. 121, here

$$e = \tilde{e}\left(\frac{\alpha}{\tilde{\alpha}}\right)\frac{1}{\sqrt{1-\tilde{m}_e/\tilde{m}_p}\sqrt{1-\tilde{\alpha}^2/\beta^2}}$$

Yields

$$(178) \qquad e = \left(\frac{48\tilde{\alpha}^2}{c^2\beta}\right)\left(\frac{\alpha}{\tilde{\alpha}}\right)\frac{1}{\sqrt{1-\tilde{m}_e/\tilde{m}_p}\sqrt{1-\tilde{\alpha}^2/\beta^2}}$$

Multiplying the right side of Eq. 178 by the expression $\frac{\sqrt{1-\tilde{m}_e/\tilde{m}_p}}{\sqrt{1-\tilde{m}_e/\tilde{m}_p}}$ yields

$$(179) \qquad e = \left[\frac{48\alpha}{c^2\left(1-\tilde{m}_e/\tilde{m}_p\right)}\left(\frac{\tilde{\alpha}}{\beta}\right)\right]\left[\frac{\sqrt{1-\tilde{m}_e/\tilde{m}_p}}{\sqrt{1-\tilde{\alpha}^2/\beta^2}}\right]$$

The second expression in brackets on the right side of Eq. 179 is the square root of Eq. 103, which relates to the correction factor μ_e/μ_B of the electron magnetic moment, here

$$\mu_e/\mu_B = \frac{(1-\tilde{m}_e/\tilde{m}_p)}{(1-\tilde{\alpha}^2/\beta^2)}$$

Substituting Eq. 103, along with the relation $\frac{\tilde{\alpha}}{\beta} = \frac{4}{\pi^4}$ from Eq. 36, into Eq. 179 yields

$$(180) \qquad e = \frac{48\alpha}{c^2\left(1-\tilde{m}_e/\tilde{m}_p\right)}\left(\frac{4}{\pi^4}\right)\sqrt{\mu_e/\mu_B}$$

Note that if we substitute the initial values for the $\tilde{m}_e/\tilde{m}_p$ ratio in Eq. 180 for the final theoretical values of the m_e/m_p ratio (this is the way it is actually used in the familiar way, which gives an insignificant difference), then Eq. 180 describes a relation between the final theoretical constants values only, which matches the experiments!

Now, substituting Eq.17a, here

$$h = m_e v_e 2\pi\, a_e$$

into Eq. 16, in the following form

$$e_{electron}^2 = m_e v_e^2 4\pi \; a_e \varepsilon_\circ$$

along with the known Maxwell's relation $\varepsilon_\circ = 1/\mu_\circ c^2$, and $v_e = c\alpha$ from Eq. 6, it yields

$$(181) \qquad \frac{e^2 c\,\mu_\circ}{2h} = \alpha$$

Substituting Eq. 181 into Eq. 180 yields

$$(182) \qquad \frac{e}{h} = \frac{\pi^4 c\,(1 - \tilde{m}_e/\tilde{m}_p)}{96\sqrt{\mu_e/\mu_B} \times \mu_\circ}$$

Note: Eq. 182 describes a relation between the final theoretical constants that matches the experiments only with the substitution of the reduced mass as mentioned for Eq. 180!

Eq. 182 represents the relation between the elementary charge e and Planck constant h, and is also half of the relation known as the Josephson frequency-voltage parameter[6] $2 \times \frac{e}{h}$, here

$$2\frac{e}{h} = \frac{\pi^4 c\,(1 - \tilde{m}_e/\tilde{m}_p)}{48\sqrt{\mu_e/\mu_B} \times \mu_\circ}$$

Substituting the following data values into the last equation

Initial value of the electron mass: $\tilde{m}_e = 9.051816380 \times 10^{-31}\,[kg]$

Initial value of the proton mass: $\tilde{m}_p = 1.663190691 \times 10^{-27}\,[kg]$

Theoretical value of the speed of light in vacuum: $c = 2.99792344 \times 10^8\,[m \; s^{-1}]$

Theoretical value of the permeability constant: $\mu_\circ = 1.26637056 \times 10^{-6}\,[H \; m^{-1}]$

Theoretical value of the correction $\sqrt{\mu_e/\mu_B}$ factor: $\sqrt{\mu_e/\mu_B} = 1.001155925$

Yields the theoretical value of the Josephson frequency-voltage parameter

$$2 \times \frac{e}{h} = 4.835948195 \times 10^{14}\,[H_Z \; V^{-1}]$$

or

$$2 \times \frac{e}{h} = 483594.8195 \times 10^{14}\,[GH_Z \; V^{-1}]$$

For comparison, the experimentally received value [8] is

$$2 \times \frac{e}{h} = 4.8359767 \times 10^{14}\,[H_Z \; V^{-1}]$$

Now, multiplying Eq. 181 by Eq. 182 yields

$$e^3 = \frac{\pi^4 h^2 \alpha\,(1 - \tilde{m}_e/\tilde{m}_p)}{48\sqrt{\mu_e/\mu_B} \times \mu_\circ^2}$$

Or after transposing and rearranging, it yields

(183) $$\left(2\frac{e}{h}\right)\left(\frac{e^2}{h}\right)=\frac{\pi^4\alpha(1-\widetilde{m}_e/\widetilde{m}_p)}{24\sqrt{\mu_e/\mu_B}\times\mu_\circ^2}$$

The left side of Eq. 183 is equal to the multiplication of the Josephson frequency-voltage parameter $2\times\frac{e}{h}$ by a constant known in physics as the Hall conductivity $\frac{e^2}{h}$.

Now, rearranging Eq. 181 (or we can place the expression $2\times\frac{e}{h}$ in Eq. 183) yields the Hall resistance R_H, that is the inverse of the Hall conductivity expression

(184) $$R_H=\frac{h}{e^2}=\frac{\mu_\circ c}{2\alpha}$$

Substituting the following data values into Eq. 184:

Theoretical value of the permeability of vacuum constant: $\mu_\circ=1.26637056\times10^{-6}\left[H\ m^{-1}\right]$

Theoretical value of the speed of light in vacuum: $c=2.99792344\times10^{8}\left[m\ s^{-1}\right]$

Theoretical value of the fine structure constant: $\alpha=\frac{1}{137.03593859}$

Yields the theoretical value of the Hall resistance

$R_H=25812.78625[\Omega]$

For comparison, the known experimental value is

$R_H=25812.8056[\Omega]$

*The electron's magnetic moment μ_e

Using Eq. 118 and multiply it by $\frac{\tilde{m}_e}{\tilde{m}_e}$ as following:

(185) $$e^2 = \frac{\tilde{m}_e v_e^2 4\pi\tilde{a}_e^2}{6\left(1-\tilde{m}_e/\tilde{m}_p\right)\left(1-\tilde{\alpha}^2/\beta^2\right)} \times \frac{\tilde{m}_e}{\tilde{m}_e} \rightarrow e^2 = \frac{\tilde{m}_e^2 v_e^2 4\pi\,\tilde{a}_e^2}{6\left(1-\tilde{m}_e/\tilde{m}_p\right)\left(1-\tilde{\alpha}^2/\beta^2\right)\tilde{m}_e}$$

Using equations Eq. 131 and Eq. 133 to express the values of $\tilde{a}_e$ and $\tilde{m}_e$ as following:

$$a_e = \tilde{a}_e\left[\frac{1}{(1-\tilde{m}_e/\tilde{m}_p)^{3/2}}\left(\frac{1}{\eta}\right)\left(\frac{\alpha^2}{\tilde{\alpha}^2}\right)\right] \rightarrow \tilde{a}_e = a_e \times (1-\tilde{m}_e/\tilde{m}_p)^{3/2} \times \eta \times \left(\frac{\tilde{\alpha}^2}{\alpha^2}\right)$$

$$m_e = \tilde{m}_e\left[\frac{(1-\tilde{m}_e/\tilde{m}_p)^{1/2}}{(1-\tilde{\alpha}^2/\beta^2)}\right]\left(\frac{\tilde{\alpha}^2}{\alpha^2}\right) \rightarrow \tilde{m}_e = m_e \times \frac{(1-\tilde{\alpha}^2/\beta^2)}{(1-\tilde{m}_e/\tilde{m}_p)^{1/2}} \times \left(\frac{\alpha^2}{\tilde{\alpha}^2}\right)$$

Multiplying $\tilde{m}_e \times \tilde{a}_e$ here:

$$\tilde{m}_e \times \tilde{a}_e = m_e \times \frac{(1-\tilde{\alpha}^2/\beta^2)}{(1-\tilde{m}_e/\tilde{m}_p)^{1/2}} \times \left(\frac{\alpha^2}{\tilde{\alpha}^2}\right) \times (1-\tilde{m}_e/\tilde{m}_p)^{3/2} \times \eta \times \left(\frac{\tilde{\alpha}^2}{\alpha^2}\right)$$

Yields:

$$\tilde{m}_e \times \tilde{a}_e = m_e \times a_e \times (1-\tilde{\alpha}^2/\beta^2) \times (1-\tilde{m}_e/\tilde{m}_p) \times \eta$$

For further development we need the multiplication of the square of each party:

$$\tilde{m}_e^2 \times \tilde{a}_e^2 = m_e^2 \times a_e^2 \times (1-\tilde{\alpha}^2/\beta^2)^2 \times (1-\tilde{m}_e/\tilde{m}_p)^2 \times \eta^2$$

Substitute the last result in Eq. 185 yields:

$$e^2 = \frac{\tilde{m}_e^2 v_e^2 4\pi\tilde{a}_e^2}{6\left(1-\tilde{m}_e/\tilde{m}_p\right)\left(1-\tilde{\alpha}^2/\beta^2\right)\tilde{m}_e} \rightarrow e^2 = \frac{m_e^2 \times v_e^2 \times 4\pi \times a_e^2 \times (1-\tilde{\alpha}^2/\beta^2)^2 \times (1-\tilde{m}_e/\tilde{m}_p)^2 \times \eta^2}{6\left(1-\tilde{m}_e/\tilde{m}_p\right)\left(1-\tilde{\alpha}^2/\beta^2\right)\tilde{m}_e}$$

After reducing:

(186) $$e^2 = \frac{m_e^2 \times v_e^2 \times 4\pi \times a_e^2 \times (1-\tilde{\alpha}^2/\beta^2) \times (1-\tilde{m}_e/\tilde{m}_p) \times \eta^2}{6\,\tilde{m}_e}$$

Using Eq. 17a here:

$$\hbar = m_e \times v_e \times a_e$$

And Substitute it in Eq. 186 yields:

(187) $$e^2 = \frac{\hbar^2 \times 4\pi \times (1-\tilde{\alpha}^2/\beta^2) \times (1-\tilde{m}_e/\tilde{m}_p) \times \eta^2}{6\,\tilde{m}_e}$$

Using the following expression obtained from substituting Eq.17a in Eq. 16 here:

$$\hbar = \frac{e^2}{2 \times v_e \times \varepsilon_\circ \times 2\pi}$$

With the expression of η constant presented at chapter **6**, which presents the ratio between the experimentally received value of the permittivity of vacuum $\varepsilon_\circ$ and its initial value $\tilde{\varepsilon}_\circ$:

$$\varepsilon_{\circ} = \tilde{\varepsilon}_{\circ} \times \eta \quad \rightarrow \hbar = \frac{e^2}{2 \times v_e \times \tilde{\varepsilon}_{\circ} \times \eta \times 2\pi}$$

And substitute it with one of the terms of $\hbar$ in Eq.187 here:

$$e^2 = \frac{\hbar \times 4\pi \times (1 - \tilde{\alpha}^2 / \beta^2) \times (1 - \tilde{m}_e / \tilde{m}_p) \times \eta^2}{6\,\tilde{m}_e} \times \frac{e^2}{2 \times v_e \times \tilde{\varepsilon}_{\circ} \times \eta \times 2\pi}$$

With reducing and substituting of $6\tilde{\varepsilon}_{\circ} = \tilde{a}_e$ from Eq. 15 it yields:

(188) $$e^2 = \hbar \times 4 \times (1 - \tilde{\alpha}^2 / \beta^2) \times (1 - \tilde{m}_e / \tilde{m}_p) \times \eta \times \frac{e^2}{2 \times \tilde{m}_e \times v_e \times \tilde{a}_e \times 2}$$

Substitute the previous result for $\tilde{m}_e \times \tilde{a}_e$ here:

$$\tilde{m}_e \times \tilde{a}_e = m_e \times a_e \times (1 - \tilde{\alpha}^2 / \beta^2) \times (1 - \tilde{m}_e / \tilde{m}_p) \times \eta$$

In Eq. 188 yields:

$$e^2 = \hbar \times 4 \times (1 - \tilde{\alpha}^2 / \beta^2) \times (1 - \tilde{m}_e / \tilde{m}_p) \times \eta \times \frac{e^2}{2 \times m_e \times v_e \times a_e \times (1 - \tilde{\alpha}^2 / \beta^2) \times (1 - \tilde{m}_e / \tilde{m}_p) \times \eta \times 2}$$

And after reducing and transposing between equation flanks it yields:

(189) $$\frac{e \times v_e \times a_e}{2} \times 2 \times \left[\frac{1 - \tilde{m}_e / \tilde{m}_p}{1 - \tilde{\alpha}^2 / \beta^2}\right] = \frac{e \times \hbar}{2 \times m_e} \times 2 \times \left[\frac{1 - \tilde{m}_e / \tilde{m}_p}{1 - \tilde{\alpha}^2 / \beta^2}\right]$$

Each sides of Eq. 189 equals to the 'electron's magnetic moment' μ_e here:

$$\mu_e = \frac{e \times \hbar}{2 \times m_e} \times 2 \times \left[\frac{1 - \tilde{m}_e / \tilde{m}_p}{1 - \tilde{\alpha}^2 / \beta^2}\right]$$

Or

$$\mu_e = \frac{e \times v_e \times a_e}{2} \times 2 \times \left[\frac{1 - \tilde{m}_e / \tilde{m}_p}{1 - \tilde{\alpha}^2 / \beta^2}\right]$$

The term in square brackets at each is the 'electron magnetic moment to Bohr magneton ratio μ_e / μ_B

(190) $$\mu_e / \mu_B = \frac{1 - \tilde{m}_e / \tilde{m}_p}{1 - \tilde{\alpha}^2 / \beta^2} = 1.001159652$$

And the electron's magnetic moment μ_e can be presented as following:

$$\mu_e = \frac{e \times \hbar}{2 \times m_e} \times 2(\mu_e / \mu_B)$$

$$\mu_e = \frac{e \times v_e \times a_e}{2} \times 2(\mu_e / \mu_B)$$

The electron magnetic moment to Bohr magneton ratio μ_e / μ_B multiplied by **2** as appears in the last equations yields the electron's Dirac factor g :

$$g = 2 \times (\mu_e / \mu_B)$$

And the electron's magnetic moment μ_e can be written as following as well (next page):

$$\mu_e = \frac{e \times \hbar}{2 \times m_e} \times g$$

And

$$\mu_e = \frac{e \times v_e \times a_e}{2} \times g$$

Where the Dirac factor g is:

$g = 2.0023193$

*The electron magnetic moment to Bohr magneton ratio μ_e/μ_B

I want to show that the value obtained by QED calculations of electron magnetic moment $\mu_e/\mu_{B\,QED} = 1.00115965221$ can be obtained by using the theoretical expression developed in Eq. 103 using the experimentally received values only!

Starting with Eq. 179, here:

$$e = \left[\frac{48\alpha}{c^2\left(1 - \tilde{m}_e/\tilde{m}_p\right)}\left(\frac{\tilde{\alpha}}{\beta}\right)\right]\left[\frac{\sqrt{1 - \tilde{m}_e/\tilde{m}_p}}{\sqrt{1 - \tilde{\alpha}^2/\beta^2}}\right]$$

And substituting Eq. 81

$$\sqrt{1 - \tilde{\alpha}^2/\beta^2} \equiv \frac{1}{192\pi^4\tilde{\alpha}^2}$$

Into the denominator of the second expression in the brackets on the right side of Eq. 179 along with the relation $\frac{\beta}{\tilde{\alpha}} = \frac{\pi^4}{4}$ from Eq. 36, yields

$$e = \frac{4 \times 48\alpha}{c^2\pi^4\left(1 - \tilde{m}_e/\tilde{m}_p\right)}\sqrt{1 - \tilde{m}_e/\tilde{m}_p}\left(192\pi^4\tilde{\alpha}^2\right)$$

Or by reducing and extracting $\tilde{\alpha}^2$

$$\tilde{\alpha}^2 = \frac{ec^2\sqrt{1 - \tilde{m}_e/\tilde{m}_p}}{192^2\alpha}$$

and after extracting the square root from both sides of the equality

$$\tilde{\alpha} = \frac{e^{1/2}c\left(1 - \tilde{m}_e/\tilde{m}_p\right)^{1/4}}{192\alpha^{1/2}}$$

Substituting the following values into the last equation for finding $\tilde{\alpha}$:

Experimentally received value of speed of light in vacuum: $c = 2.99792458 \times 10^8 \left[m\ s^{-1}\right]$

Experimentally received value of the elementary charge: $e = 1.6021892 \times 10^{-19}\left[C\right]$

Experimentally received value of the proton mass: $m_p = 1.6726485 \times 10^{-27}\left[Kg\right]$

Experimentally received value of the proton mass: $m_e = 9.109534 \times 10^{-31}\left[Kg\right]$

Reduced mass factor based on experimental values: $\dfrac{m_p}{m_e + m_p} = 0.999455679$

Obtained Experimental value for the fine structure constant: $\alpha = \dfrac{1}{137.035996} = 7.29735272 \times 10^{-3}$

Yields the initial fine structure constant $\tilde{\alpha}$ calculated by the experimentally received values!

$\tilde{\alpha} = 7.31534124\,6 \times 10^{-3}$

For comparison, the value calculated by this theory is:

$\tilde{\alpha} = 7.31533429\,6 \times 10^{-3}$

Substituting $\tilde{\alpha}$ calculated by the experimentally received values into Eq. 81, yields

$$\sqrt{1 - \tilde{\alpha}^2/\beta^2} \equiv \frac{1}{192\pi^4\tilde{\alpha}^2} = 0.999148638 \rightarrow (1 - \tilde{\alpha}^2/\beta^2) = 0.998298$$

The reduced mass, as mentioned for Eq. 180 earlier meaning:

$$1 - \frac{\tilde{m}_e}{\tilde{m}_p} = 0.999455756 \text{ (from theory)} \cong \frac{m_p}{m_e + m_p} = 0.999455679 \text{ (experimentally received)}$$

Substituting the last two solutions into the theoretical expression developed in Eq. 103, yields the Electron magnetic moment μ_e/μ_B factor based on experimentally received values only!

$$\mu_e/\mu_B = \frac{\left(\dfrac{m_p}{m_e + m_p}\right)}{\left(1 - \dfrac{\tilde{\alpha}^2}{\beta^2}\right)} = \frac{0.999455679}{0.998298} = 1.001159652$$

The value of the Electron magnetic moment factor μ_e/μ_B calculated by using the QED methods is:

$\mu_e/\mu_{B\,QED} = 1.00115965\,221$

This implies that the source for this value is the theoretical expression developed in Eq. 103. Therefore the exact value of the Electron magnetic moment factor μ_e/μ_B should be based on the initial values calculated by this theory:

The initial fine structure constant: $\tilde{\alpha} = 7.31534183\,7 \times 10^{-3}$

The reduced mass $(1 - \tilde{m}_e/\tilde{m}_p) = 0.999455756$

That yields

$\mu_e/\mu_B = 1.001155925$

3

The Proton Magnetic Moment to Nuclear magneton ratio μ_p/μ_N

We start by using Eq. 23 here with: $c\times\beta = v_p$

(3.a) $\tilde{e}^2 = \dfrac{\tilde{m}_p v_p^2 4\pi\tilde{r}_p^2(1-\beta^2)}{6}\times\dfrac{\tilde{m}_p}{\tilde{m}_e}$

To transform the initial value of the electric charge $\tilde{e}$ to its relativistic one e, we use Eq. 121 here:

$e^2 = \tilde{e}^2\times\left(\dfrac{\alpha^2}{\tilde{\alpha}^2}\right)\times\dfrac{1}{(1-\tilde{m}_e/\tilde{m}_p)(1-\tilde{\alpha}^2/\beta^2)}$, it yields:

(3.b) $e^2 = \dfrac{\tilde{m}_p v_p^2 4\pi\tilde{r}_p^2(1-\beta^2)}{6}\times\dfrac{\tilde{m}_p}{\tilde{m}_e}\times\left(\dfrac{\alpha^2}{\tilde{\alpha}^2}\right)\times\dfrac{1}{(1-\tilde{m}_e/\tilde{m}_p)(1-\tilde{\alpha}^2/\beta^2)}$

For the next step we need to find the relation for $\tilde{m}_p\tilde{r}_p$ thru multiplying of Eq. 147 by Eq. 142 here:

$\tilde{m}_p = m_p\times\left[\dfrac{\tilde{m}_e(1-\tilde{m}_e/\tilde{m}_p)}{m_e}\right]\times\sqrt{\dfrac{\tilde{\alpha}}{\alpha}}$ and $\rightarrow$ $\tilde{r}_p = r_p\times\left[\dfrac{\tilde{a}_e}{a_e(1-\tilde{m}_e/\tilde{m}_p)}\right]\times\sqrt{\dfrac{\tilde{\alpha}}{\alpha}}$, it yields:

(3.c) $\tilde{m}_p\tilde{r}_p = m_p r_p\times\left(\dfrac{\tilde{m}_e\tilde{a}_e}{m_e a_e}\right)\times\dfrac{\tilde{\alpha}}{\alpha}$

To find the expression for: $\left(\dfrac{\tilde{m}_e\tilde{a}_e}{m_e a_e}\right)$, we multiply Eq. 131 by Eq. 133 here:

$\tilde{a}_e = a_e\times(1-\tilde{m}_e/\tilde{m}_p)^{3/2}\times\eta\times\left(\dfrac{\tilde{\alpha}^2}{\alpha^2}\right)$ and $\rightarrow$ $\tilde{m}_e = m_e\times\left[\dfrac{(1-\tilde{\alpha}^2/\beta^2)}{(1-\tilde{m}_e/\tilde{m}_p)^{1/2}}\right]\times\left(\dfrac{\alpha^2}{\tilde{\alpha}^2}\right)$, it yields:

(3.d) $\dfrac{\tilde{m}_e\tilde{a}_e}{m_e a_e} = (1-\tilde{m}_e/\tilde{m}_p)\times(1-\tilde{\alpha}^2/\beta^2)\times\eta$

Substitute the result of **(3.d)** in **(3.c),** yields:

$\tilde{m}_p\tilde{r}_p = m_p r_p\times(1-\tilde{m}_e/\tilde{m}_p)\times(1-\tilde{\alpha}^2/\beta^2)\times\eta\times\dfrac{\tilde{\alpha}}{\alpha}$, or the square of it following here:

(3.e) $\tilde{m}_p^2\tilde{r}_p^2 = m_p^2 r_p^2\times(1-\tilde{m}_e/\tilde{m}_p)^2\times(1-\tilde{\alpha}^2/\beta^2)^2\times\eta^2\times\dfrac{\tilde{\alpha}^2}{\alpha^2}$

Substitute the result of **(3.e)** in **(3.b),** yields:

$$e^2 = m_p v_p r_p\sqrt{1-\beta^2}\times m_p v_p r_p\sqrt{1-\beta^2}\,\frac{4\pi}{6\tilde{m}_e}\times\left(\frac{\alpha^2}{\tilde{\alpha}^2}\right)\times\frac{(1-\tilde{m}_e/\tilde{m}_p)^2\times(1-\tilde{\alpha}^2/\beta^2)^2}{(1-\tilde{m}_e/\tilde{m}_p)(1-\tilde{\alpha}^2/\beta^2)}\times\eta^2\times\frac{\tilde{\alpha}^2}{\alpha^2}$$

Substitute the expression: $\hbar = m_p v_p r_p\sqrt{1-\beta^2}$ from Eq. 17b in it, yields:

(3.f) $e^2 = \hbar^2\times\dfrac{4\pi}{6\tilde{m}_e}\times\left(\dfrac{\alpha^2}{\tilde{\alpha}^2}\right)\times\dfrac{(1-\tilde{m}_e/\tilde{m}_p)^2\times(1-\tilde{\alpha}^2/\beta^2)^2}{(1-\tilde{m}_e/\tilde{m}_p)(1-\tilde{\alpha}^2/\beta^2)}\times\eta^2\times\dfrac{\tilde{\alpha}^2}{\alpha^2}$

Substitute Eq. 16 combined with Eq. 17a: $\hbar = \dfrac{e^2}{2v_e\varepsilon_\circ 2\pi}$ along with $\varepsilon_\circ = \tilde{\varepsilon}_\circ\times\eta$ from page 41 in yields:

(3.g) $\hbar = \dfrac{e^2}{2\times v_e\times\tilde{\varepsilon}_\circ\times\eta\times 2\pi}$

Substitute **(3.g)** in the right flank of **(3.f)**, replacing one of $\hbar$ at the numerator, yields:

$$e^2 = \frac{e^2}{2 \times v_e \times 6\tilde{\varepsilon}_\circ \times \eta \times 2\pi} \times \hbar \times \frac{4\pi}{\tilde{m}_e} \times \left(\frac{\alpha^2}{\tilde{\alpha}^2}\right) \times \frac{(1-\tilde{m}_e/\tilde{m}_p)^2 \times (1-\tilde{\alpha}^2/\beta^2)^2}{(1-\tilde{m}_e/\tilde{m}_p)(1-\tilde{\alpha}^2/\beta^2)} \times \eta^2 \times \frac{\tilde{\alpha}^2}{\alpha^2}$$

Substitute $6\tilde{\varepsilon}_\circ = \tilde{a}_e$ from Eq. 15 in it, yields:

(3.h) $$e^2 = \frac{e^2}{2 \times \tilde{m}_e \tilde{a}_e \times v_e \times \eta} \times \hbar \times \frac{4\pi}{2\pi} \times \left(\frac{\alpha^2}{\tilde{\alpha}^2}\right) \times \frac{(1-\tilde{m}_e/\tilde{m}_p)^2 \times (1-\tilde{\alpha}^2/\beta^2)^2}{(1-\tilde{m}_e/\tilde{m}_p)(1-\tilde{\alpha}^2/\beta^2)} \times \eta^2 \times \frac{\tilde{\alpha}^2}{\alpha^2}$$

Substitute the following relation from **(3.d)** in **(3.h)**:

$\tilde{m}_e \tilde{a}_e = m_e a_e \times (1-\tilde{m}_e/\tilde{m}_p) \times (1-\tilde{\alpha}^2/\beta^2) \times \eta$, it yields:

(3.i) $$e^2 = \frac{e^2}{2 \times m_e a_e v_e \times \eta^2} \times \hbar \times \frac{4\pi}{2\pi} \times \left(\frac{\alpha^2}{\tilde{\alpha}^2}\right) \times \frac{(1-\tilde{m}_e/\tilde{m}_p)^2 \times (1-\tilde{\alpha}^2/\beta^2)^2}{(1-\tilde{m}_e/\tilde{m}_p)^2(1-\tilde{\alpha}^2/\beta^2)^2} \times \eta^2 \times \frac{\tilde{\alpha}^2}{\alpha^2}$$

Substitute: $m_e v_e a_e = m_p v_p r_p \sqrt{1-\beta^2}$ from Eq. 20 in **(3.i)**, yields:

(3.j) $$e^2 = \frac{e^2}{2 \times m_p v_p r_p \sqrt{1-\beta^2} \times \eta^2} \times \hbar \times \frac{4}{2} \times \left(\frac{\alpha^2}{\tilde{\alpha}^2}\right) \times \frac{(1-\tilde{m}_e/\tilde{m}_p)^2 \times (1-\tilde{\alpha}^2/\beta^2)^2}{(1-\tilde{m}_e/\tilde{m}_p)^2(1-\tilde{\alpha}^2/\beta^2)^2} \times \eta^2 \times \frac{\tilde{\alpha}^2}{\alpha^2}$$

Reducing and transposing between flanks, yields:

(3.k) $$\frac{e\, v_p r_p}{2} \times \left(\frac{1}{2} \times \frac{\alpha^2}{\tilde{\alpha}^2}\right) = \frac{e\hbar}{2m_p\sqrt{1-\beta^2}} \times \left(\frac{1}{2} \times \frac{\alpha^2}{\tilde{\alpha}^2}\right)$$

Substitute: $\sqrt{1-\beta^2} = \pi^3\beta^2$ from Eq. 37 in **(3.k),** yields:

(3.l) $$\frac{e\, v_p r_p}{2} \times \left(\frac{1}{2} \times \frac{\alpha^2}{\tilde{\alpha}^2}\right) = \frac{e\hbar}{2m_p\pi^3\beta^2} \times \left(\frac{1}{2} \times \frac{\alpha^2}{\tilde{\alpha}^2}\right)$$

Multiplying both flanks by ($\pi^3\beta$) which appears at the denominator of the right flank, yields:

(3.m) $$\frac{e\, v_p r_p}{2} \times \left(\frac{1}{2} \times \frac{\alpha^2}{\tilde{\alpha}^2}\right) \times \pi^3\beta = \frac{e\hbar}{2m_p} \times \left(\frac{1}{2\beta} \times \frac{\alpha^2}{\tilde{\alpha}^2}\right)$$

Substitute: $\frac{\sqrt{1-\beta^2}}{\beta} = \pi^3\beta$ from Eq. 37 in the right flank of **(3.m)**, yields:

(3.n) $$\frac{e\, v_p r_p \sqrt{1-\beta^2}}{2} \times \left(\frac{1}{2\beta} \times \frac{\alpha^2}{\tilde{\alpha}^2}\right) = \frac{e\hbar}{2m_p} \times \left(\frac{1}{2\beta} \times \frac{\alpha^2}{\tilde{\alpha}^2}\right)$$

Both flanks of equation **(3.n)** are the expressions for the 'Proton Magnetic Moment' μ_p.

$$\mu_p = \frac{e\hbar}{2m_p} \times \left(\frac{1}{2\beta} \times \frac{\alpha^2}{\tilde{\alpha}^2}\right) \quad \text{Or} \quad \mu_p = \frac{e\, v_p r_p \sqrt{1-\beta^2}}{2} \times \left(\frac{1}{2\beta} \times \frac{\alpha^2}{\tilde{\alpha}^2}\right)$$

And the 'Proton Magnetic Moment to Nuclear magneton ratio' μ_p/μ_N is the term in the parenthesis:

(3.o) $$\mu_p/\mu_N = \frac{1}{2\beta} \times \frac{\alpha^2}{\tilde{\alpha}^2}$$

Note: In the 'Three Energy Levels' model presented from 'The Quarks', the integer **2** at the μ_p/μ_N denominator, indicates the second energy level which is the dominant energy level in the proton due to the presence of the two *up* quarks at this level, which creates the effect of the 'positive magnetic moment' of the proton that partly is influenced by interaction with the orbiting electrons in the atoms.

Substitute the values of α, $\tilde{\alpha}$ and β in **(3.o)**, yields: $\mu_p/\mu_N = 2.7929233$

For comparison, the experimentally received value [8] is: $\mu_p/\mu_N = 2.792847386$

4

The Neutron Magnetic Moment to Nuclear Magneton ratio μ_n/μ_N

The first part of the process of finding the expression for μ_n/μ_N is completely compatible to that developed for the proton's μ_p/μ_N expression in the previous chapter, in context of the equations used and the stages of development up to paragraph **(3.i)**. Please read first stages of development done for finding μ_p/μ_N in the previous chapter up to paragraph **(3.i)** before you continue reading here.

First I will write here equation of **(3.i)** that will be used further on.

$$e^2 = \frac{e^2}{2\times m_e a_e v_e \times \eta^2} \times \hbar \times \frac{4\pi}{2\pi} \times \left(\frac{\alpha^2}{\tilde{\alpha}^2}\right) \times \frac{(1-\tilde{m}_e/\tilde{m}_p)^2 \times (1-\tilde{\alpha}^2/\beta^2)^2}{(1-\tilde{m}_e/\tilde{m}_p)^2 (1-\tilde{\alpha}^2/\beta^2)^2} \times \eta^2 \times \frac{\tilde{\alpha}^2}{\alpha^2}$$

First we need to find the relation for $\tilde{m}_p\tilde{r}_p$ thru equating Eq. 32 and Eq. 34 Here:

(4.a) $$\left(\frac{\tilde{m}_p}{\tilde{m}_e}\right)\frac{4\tilde{\alpha}\beta\sqrt{1-\beta^2}}{3\pi} = \left(\frac{\tilde{a}_e}{\tilde{r}_p}\right)\frac{\tilde{\alpha}^3\sqrt{1-\beta^2}}{3\beta^3}$$

By transposing flanks it yields:

(4.b) $$\left(\frac{\tilde{m}_p\tilde{r}_p}{\tilde{m}_e\tilde{a}_e}\right) = \frac{3\pi\,\tilde{\alpha}^2\sqrt{1-\beta^2}}{4\beta\sqrt{1-\beta^2}\times 3\beta^3}$$

Using the relation found for $\tilde{m}_p\tilde{r}_p$ at clause **(3.e)** from previous chapter 3, here:

$$\tilde{m}_p\tilde{r}_p = m_p r_p \times (1-\tilde{m}_e/\tilde{m}_p)\times(1-\tilde{\alpha}^2/\beta^2)\times\eta\times\frac{\tilde{\alpha}}{\alpha}$$

Along with the relation found for $\tilde{m}_e\tilde{a}_e$ at clause **(3.h)** from previous chapter 3, here:

$$\tilde{m}_e\tilde{a}_e = m_e a_e(1-\tilde{m}_e/\tilde{m}_p)\times(1-\tilde{\alpha}^2/\beta^2)\times\eta$$

To find the expression for $\left(\frac{\tilde{m}_p\tilde{r}_p}{\tilde{m}_e\tilde{a}_e}\right)$ by dividing the first from the second yields:

(4.c) $$\left(\frac{\tilde{m}_p\tilde{r}_p}{\tilde{m}_e\tilde{a}_e}\right) = \frac{m_p r_p \times(1-\tilde{m}_e/\tilde{m}_p)\times(1-\tilde{\alpha}^2/\beta^2)\times\eta\times\frac{\tilde{\alpha}}{\alpha}}{m_e a_e(1-\tilde{m}_e/\tilde{m}_p)\times(1-\tilde{\alpha}^2/\beta^2)\times\eta} \rightarrow \left(\frac{\tilde{m}_p\tilde{r}_p}{\tilde{m}_e\tilde{a}_e}\right) = \frac{m_p r_p\tilde{\alpha}}{m_e a_e\alpha}$$

Equating **(4.c)** and **(4.b)** yields:

(4.d) $$\frac{m_p r_p\tilde{\alpha}}{m_e a_e\alpha} = \frac{3\pi\,\tilde{\alpha}^2\sqrt{1-\beta^2}}{4\beta\sqrt{1-\beta^2}\times 3\beta^3} \rightarrow \frac{m_p r_p\beta\sqrt{1-\beta^2}}{m_e a_e\alpha} = \frac{3\pi\,\tilde{\alpha}\sqrt{1-\beta^2}}{4\times 3\beta^3}$$

Substitute $\beta^2 = \frac{\sqrt{1-\beta^2}}{\pi^3}$ from Eq. 37 at the right flank denominator of **(4.d)** without reducing, yields:

$$\frac{m_p r_p\beta\sqrt{1-\beta^2}}{m_e a_e\alpha} = \frac{3\pi\,\tilde{\alpha}\sqrt{1-\beta^2}}{4\times 3\beta\times\beta^2} \rightarrow \frac{m_p r_p\beta\sqrt{1-\beta^2}}{m_e a_e\alpha} = \frac{3\pi^4\,\tilde{\alpha}\sqrt{1-\beta^2}}{3\times 4\beta\sqrt{1-\beta^2}}$$

And substitute $\frac{\pi^4}{4} = \frac{\beta}{\tilde{\alpha}}$ from Eq. 36 at the right flank, yields:

(4.e) $$\frac{m_p r_p \beta\sqrt{1-\beta^2}}{m_e a_e \alpha} = \frac{3\pi\,\tilde{\alpha}\sqrt{1-\beta^2}}{4\times 3\beta\times\beta^2} \rightarrow \frac{m_p r_p \beta\sqrt{1-\beta^2}}{m_e a_e \alpha} = \frac{3\beta\sqrt{1-\beta^2}}{3\beta\sqrt{1-\beta^2}}$$

Substituting: $m_p r_p = m_n r_n$ from Eq. 171 and multiplying nominator and denominator of **(4.e)** left flank by the speed of light in vacuum c for setting the velocities: $c\alpha = v_e$ and $c\beta = v_p$ (the velocity v_p is the same for the neutron as for the proton), yields:

(4.f) $$\frac{m_n r_n v_p\sqrt{1-\beta^2}}{m_e a_e v_e} = \frac{3\beta\sqrt{1-\beta^2}}{3\beta\sqrt{1-\beta^2}} \quad \text{or} \rightarrow \quad m_e a_e v_e = \frac{m_n r_n v_p\sqrt{1-\beta^2}\times 3\beta\sqrt{1-\beta^2}}{3\beta\sqrt{1-\beta^2}}$$

Substituting **(4.f)** in **(3.i)** (the equation from the previous chapter), it yields:

$$e^2 = \frac{e^2\times 3\beta\sqrt{1-\beta^2}}{2\times m_n r_n v_p\sqrt{1-\beta^2}\times 3\beta\sqrt{1-\beta^2}\times\eta^2}\times\hbar\times\frac{4\pi}{2\pi}\times\left(\frac{\alpha^2}{\tilde{\alpha}^2}\right)\times\frac{(1-\tilde{m}_e/\tilde{m}_p)^2\times(1-\tilde{\alpha}^2/\beta^2)^2}{(1-\tilde{m}_e/\tilde{m}_p)^2(1-\tilde{\alpha}^2/\beta^2)^2}\times\eta^2\times\frac{\tilde{\alpha}^2}{\alpha^2}$$

After reducing and transposing between flanks it yields:

(4.h) $$\frac{e\, r_n v_p\sqrt{1-\beta^2}}{2}\times\left[\frac{1}{3\beta\sqrt{1-\beta^2}\times(1-\tilde{m}_e/\tilde{m}_p)^2}\times\left(\frac{\tilde{\alpha}^2}{\alpha^2}\right)\right] = \frac{e\hbar}{2m_n}\left[\frac{1}{3\beta\sqrt{1-\beta^2}\times(1-\tilde{m}_e/\tilde{m}_p)^2}\times\left(\frac{\tilde{\alpha}^2}{\alpha^2}\right)\right]$$

Both flanks of equation **(4.h)** are the expressions for the Neutron Magnetic Moment μ_n.

$$\mu_n = \frac{e\hbar}{2m_n}\left[\frac{1}{3\beta\sqrt{1-\beta^2}\times(1-\tilde{m}_e/\tilde{m}_p)^2}\times\left(\frac{\tilde{\alpha}^2}{\alpha^2}\right)\right]$$

Or

$$\mu_n = \frac{e\, r_n v_p\sqrt{1-\beta^2}}{2}\times\left[\frac{1}{3\beta\sqrt{1-\beta^2}\times(1-\tilde{m}_e/\tilde{m}_p)^2}\times\left(\frac{\tilde{\alpha}^2}{\alpha^2}\right)\right]$$

And the 'Neutron Magnetic Moment to Nuclear magneton ratio' μ_n/μ_N is term in square brackets:

(4.i) $$\mu_n/\mu_N = \frac{1}{3\beta\sqrt{1-\beta^2}\times(1-\tilde{m}_e/\tilde{m}_p)^2}\times\left(\frac{\tilde{\alpha}^2}{\alpha^2}\right)$$

Note: In the 'Three Energy Levels' model presented from chapter 13 (The Quarks), the integer **3** at the expression of μ_n/μ_N at its denominator, is indicating the third energy level which is the dominant energy level in an 'excited state' of a neutron due to the presence of the negative Pi mesons at this level (before neutron's decaying process), which creates the effect of the 'negative magnetic moment' of the neutron, with influence from the interaction with the orbiting electrons in atoms.
The 'Neutron Magnetic Moment to Nuclear Magneton ratio' μ_n/μ_N value is the following:
Substituting the following data values in **(4.i)**:

Reduced mass factor: $(1-\tilde{m}_e/\tilde{m}_p) = 0.999455756$; The ratio of $\frac{\tilde{\alpha}^2}{\alpha^2} = 1.004933474$

Key constant: $\beta = 0.178145016$ and factor $\gamma = \sqrt{1-\beta^2} = 0.9840042445$

Yields the theoretical value of the 'Neutron Magnetic Moment to Nuclear Magneton ratio' μ_n/μ_N :
$\mu_n/\mu_N = 1.913014502$

For comparison, the experimentally value is: $\mu_n/\mu_N = 1.91304275$

5
The Avogadro Constant N_A

First, substituting the relation $\tilde{a}_e = 6\tilde{\varepsilon}_\circ$ from Eq. 15 into Eq. 22 yields

$$\tilde{e}^2 = \frac{\tilde{m}_p v_p^2 4\pi\, \tilde{r}_p \sqrt{1-\beta^2}\,\tilde{a}_e}{2\times 3}\left(\frac{\tilde{\alpha}}{\beta}\right)$$

Then substituting Eq. 22 into Eq. 121 after raising the last to the power of **2**

$$e^2 = \tilde{e}^2\left(\frac{\alpha^2}{\tilde{\alpha}^2}\right)\frac{1}{\left(1-\tilde{m}_e/\tilde{m}_p\right)\left(1-\tilde{\alpha}^2/\beta^2\right)}$$

yields the expression for the value of the theoretical elementary charge e that matches the experiments, based on the proton components

$$e^2 = \frac{\tilde{m}_p v_p^2 4\pi\, \tilde{r}_p \sqrt{1-\beta^2}\,\tilde{a}_e}{2\times 3}\left(\frac{\tilde{\alpha}}{\beta}\right)\left[\left(\frac{\alpha^2}{\tilde{\alpha}^2}\right)\frac{1}{\left(1-\tilde{m}_e/\tilde{m}_p\right)\left(1-\tilde{\alpha}^2/\beta^2\right)}\right]$$

Or, after being rearranged,

$$(191)\qquad e^2 = \left[\frac{\tilde{m}_p}{\sqrt{1-\tilde{\alpha}^2/\beta^2}}\left(\frac{\alpha}{\tilde{\alpha}}\right)\right]\frac{v_p^2 4\pi\, \tilde{r}_p \sqrt{1-\beta^2}\,\tilde{a}_e}{2\times 3}\left(\frac{\tilde{\alpha}}{\beta}\right)\left[\left(\frac{\alpha}{\tilde{\alpha}}\right)\frac{1}{\left(1-\tilde{m}_e/\tilde{m}_p\right)\sqrt{1-\tilde{\alpha}^2/\beta^2}}\right]$$

The first term in brackets on the right side of Eq. 191 is the definition for the mass number m_u

$$(192)\qquad m_u = \frac{\tilde{m}_p}{\sqrt{1-\tilde{\alpha}^2/\beta^2}}\left(\frac{\alpha}{\tilde{\alpha}}\right)$$

Substituting the following data values into Eq. 192
Initial fine structure constant: $\tilde{\alpha} = 7.315341837\times 10^{-3}$

Theoretical value of the fine structure constant: $\alpha = \dfrac{1}{137.03593859}$

Effective value of γ factor: $\sqrt{1-\tilde{\alpha}^2/\beta^2} = \dfrac{1}{192\pi^4\tilde{\alpha}^2} = 0.999150536$

Initial value of the proton mass: $\tilde{m}_p = 1.663190691 \times 10^{-27}[kg]$
Yields the theoretical value of the *mass number* m_u

$$m_u = 1.660513232 \times 10^{-27}[kg]$$

Transposing the terms in Eq. 191 yields

$$(193) \qquad \left[\frac{\sqrt{1 - \tilde{\alpha}^2/\beta^2}}{\tilde{m}_p} \times \frac{\tilde{\alpha}}{\alpha}\right] = \left(\frac{1}{e^2}\right)\frac{v_p^2 4\pi\,\tilde{r}_p\sqrt{1-\beta^2}\,\tilde{a}_e}{2\times 3}\left(\frac{\tilde{\alpha}}{\beta}\right)\left[\left(\frac{\alpha}{\tilde{\alpha}}\right)\frac{1}{\left(1-\tilde{m}_e/\tilde{m}_p\right)\sqrt{1 - \tilde{\alpha}^2/\beta^2}}\right]$$

The expression on the left side of Eq. 193 is in fact the inverse of the mass number defined at Eq. 192, and should be the definition for *Avogadro constant* N_A

$$(194) \qquad N_A = \frac{1}{m_u} = \frac{\sqrt{1 - \tilde{\alpha}^2/\beta^2}}{\tilde{m}_p}\left(\frac{\tilde{\alpha}}{\alpha}\right)$$

Based on the theoretical value received for m_u the theoretical value of the *Avogadro constant* is

$$N_A = 6.022234457 \times 10^{26}[kmol^{-1}]$$

The constant shown here represents *the quantity of protons in one kilomole*!

For comparison, the experimentally received value [2] is

$$N_A = 6.022045 \times 10^{26}[kmol^{-1}]$$

6
The Faraday Constant F

Multiply both flanks of Eq. 193 by e yields:

$$(195) \qquad \left[e \times \frac{\sqrt{1 - \tilde{\alpha}^2/\beta^2}}{\tilde{m}_p} \left(\frac{\tilde{\alpha}}{\alpha} \right) \right] = \left(\frac{1}{e} \right) \times \frac{v_p^2 4\pi\, \tilde{r}_p \sqrt{1 - \beta^2 \tilde{a}_e}}{2 \times 3} \left(\frac{\tilde{\alpha}}{\beta} \right) \left[\left(\frac{\alpha}{\tilde{\alpha}} \right) \frac{1}{\left(1 - \tilde{m}_e/\tilde{m}_p\right) \sqrt{1 - \tilde{\alpha}^2/\beta^2}} \right]$$

The term in the brackets at the left flank of Eq. 195 consist of the term $\left[\frac{\sqrt{1 - \tilde{\alpha}^2/\beta^2}}{\tilde{m}_p} \left(\frac{\tilde{\alpha}}{\alpha} \right) \right]$ which was defined as being the Avogadro constant N_A in the previous chapter. Multiplication of the Avogadro constant N_A by the elementary charge e yields the Faraday constant F

$$(196) \qquad F = e \times N_A = e \times \frac{\sqrt{1 - \tilde{\alpha}^2/\beta^2}}{\tilde{m}_p} \left(\frac{\tilde{\alpha}}{\alpha} \right)$$

Now the elementary charge e can be obtained from Eq. 121:

$$e = \tilde{e} \times \left(\frac{\alpha}{\tilde{\alpha}} \right) \frac{1}{\sqrt{1 - \tilde{m}_e / \tilde{m}_p} \sqrt{1 - \tilde{\alpha}^2 / \beta^2}}$$

Substitute the last term for the elementary charge e in Eq. 196, yields:

$$(196a) \qquad F = \frac{\tilde{e}}{\tilde{m}_p} \times \frac{1}{\sqrt{1 - \tilde{m}_e / \tilde{m}_p}}$$

Substitute the following data received from this theory:
Theoretical initial value of the elementary charge: $\tilde{e} = 1.60433419 \times 10^{-19} [C]$
Theoretical initial value of the proton mass: $\tilde{m}_p = 1.663190691 \times 10^{-27} [kg]$
Reduced mass factor: $1 - \tilde{m}_e / \tilde{m}_p = 0.99945567$
Yields the theoretical value of the Faraday constant F
$F = 96487.5 [C\,mol^{-1}]$
For comparison, the experimentally received value [2] is
$F = 96484.56 [C\,mol^{-1}]$

7
The absolute Temperature T_o the Boltzmann constant k And the Molar Gas constant R

Starting with Eq. 118:

$$e^2 = \frac{\tilde{m}_e v_e^2 4\pi \tilde{a}_e^2}{6\left(1 - \tilde{\alpha}^2 / \beta^2\right)\left(1 - \tilde{m}_e / \tilde{m}_p\right)}$$

Divide both flanks by $4\pi\tilde{a}_e^2$:

$$\frac{e^2}{4\pi \tilde{a}_e^2} = \frac{\tilde{m}_e v_e^2}{6\left(1 - \tilde{\alpha}^2 / \beta^2\right)\left(1 - \tilde{m}_e / \tilde{m}_p\right)}$$

Now substitute the term of Eq. 133 for $\tilde{m}_e$ at the right flank:

$$\tilde{m}_e = m_e\left(\frac{\alpha^2}{\tilde{\alpha}^2}\right)\left[\frac{(1 - \tilde{\alpha}^2 / \beta^2)}{(1 - \tilde{m}_e / \tilde{m}_p)^{1/2}}\right]$$, It yields:

$$\frac{e^2}{4\pi \tilde{a}_e^2} = \frac{m_e v_e^2 \alpha^2 \left(1 - \tilde{\alpha}^2 / \beta^2\right)}{6\left(1 - \tilde{\alpha}^2 / \beta^2\right)\left(1 - \tilde{m}_e / \tilde{m}_p\right)^{3/2} \tilde{\alpha}^2}$$

Substitute Eq. 131 in the left flank denominator of the last equation for $\tilde{a}_e^2$:

$$\tilde{a}_e^2 = \frac{a_e^2 \eta^2 \left(1 - \tilde{m}_e / \tilde{m}_p\right)^3 \tilde{\alpha}^4}{\alpha^4}$$

It forms:

$$\frac{e^2 \alpha^4}{4\pi a_e^2 \eta^2 \left(1 - \tilde{m}_e / \tilde{m}_p\right)^3 \tilde{\alpha}^4} = \frac{m_e v_e^2 \alpha^2 \left(1 - \tilde{\alpha}^2 / \beta^2\right)}{6\left(1 - \tilde{\alpha}^2 / \beta^2\right)\left(1 - \tilde{m}_e / \tilde{m}_p\right)^{3/2} \tilde{\alpha}^2}$$

Multiply both flanks by $\left[\frac{\eta^2}{\left(1 - \tilde{\alpha}^2 / \beta^2\right)}\right]$, yields:

(197) $$\frac{e^2 \alpha^4}{4\pi a_e^2 \left(1 - \tilde{\alpha}^2 / \beta^2\right)\left(1 - \tilde{m}_e / \tilde{m}_p\right)^3 \tilde{\alpha}^4} = \frac{m_e v_e^2 \alpha^2 \eta^2}{6\left(1 - \tilde{\alpha}^2 / \beta^2\right)\left(1 - \tilde{m}_e / \tilde{m}_p\right)^{3/2} \tilde{\alpha}^2}$$

Now the effective value of the factor γ is calculated by the expression mentioned at Eq. 100, here:

$$\gamma_{proton / electron} = \frac{1}{192\pi^4 \tilde{\alpha}^2} \equiv \sqrt{1 - \tilde{\alpha}^2 / \beta^2} \rightarrow \frac{1}{\sqrt{1 - \tilde{\alpha}^2 / \beta^2}} = 192\pi^4 \tilde{\alpha}^2$$

Substitute one of the terms $\frac{1}{\sqrt{1 - \tilde{\alpha}^2 / \beta^2}} = 768\beta\tilde{\alpha}$ at both flanks of Eq. 197, along with reducing the (768) component at both flanks and rearrange, it yields:

(198) $$\frac{e^2 \alpha^4 \beta}{2\pi a_e^2 \left(1 - \tilde{m}_e / \tilde{m}_p\right)^3 \tilde{\alpha}^4} \times \left[\frac{\tilde{\alpha}}{2\sqrt{1 - \tilde{\alpha}^2 / \beta^2}}\right] = \frac{m_e v_e^2 \alpha^2 \eta^2 \beta}{3\left(1 - \tilde{m}_e / \tilde{m}_p\right)^{3/2} \tilde{\alpha}^2} \times \left[\frac{\tilde{\alpha}}{2\sqrt{1 - \tilde{\alpha}^2 / \beta^2}}\right]$$

The term in the square brackets at each flank of Eq. 198 expresses the inverse of the absolute temperature T_o in Kelvin scale.

7-1. The absolute Temperature $T_\circ$ (Kelvin scale)

As was defined the last term at each flank of Eq. 198 is the inverse of the absolute temperature $T_\circ$ in Kelvin scale, here:

$$(199)\qquad \frac{1}{T_\circ} = \frac{\tilde{\alpha}}{2\sqrt{1-\tilde{\alpha}^2/\beta^2}} \quad \text{Or} \quad \frac{1}{T_\circ} = \frac{\tilde{v}_e}{2c\sqrt{1-\tilde{\alpha}^2/\beta^2}} \rightarrow T_\circ = \frac{1}{96\pi^4\tilde{\alpha}^3}$$

I've substituted the ratio $\tilde{\alpha} = \frac{\tilde{v}_e}{c}$ to emphasize the relation to the velocity of electrons in the atoms!

Substitute the following data values into Eq. 199:

Initial value of the fine structure constant: $\tilde{\alpha} = 7.315334297\times10^{-3}$

Yields the theoretical value of the absolute temperature $T_\circ$ in Kelvin scale:

$$\frac{1}{T_\circ} = \frac{1}{273.1660634}[1/K^\circ] \rightarrow T_\circ = 273.1660634\ [K^\circ]$$

For comparison, the experimentally received value [3] is:

$$\frac{1}{T_\circ} = \frac{1}{273.15}[1/K^\circ] \rightarrow T_\circ = 273.15[K^\circ]$$

7-2. The Boltzmann constant k

Substitute $\frac{1}{T_\circ} = \frac{\tilde{\alpha}}{2\sqrt{1-\tilde{\alpha}^2/\beta^2}}$ in Eq. 198, yields:

$$\frac{e^2\alpha^4\beta}{2\pi a_e^2\left(1-\tilde{m}_e/\tilde{m}_p\right)^3\tilde{\alpha}^4} \times \frac{1}{T_\circ} = \frac{m_e v_e^2\alpha^2\eta^2\beta}{3\left(1-\tilde{m}_e/\tilde{m}_p\right)^{3/2}\tilde{\alpha}^2} \times \frac{1}{T_\circ}$$

Now substitute $\beta = \tilde{\alpha}\pi^4/4$ from Eq. 36 and after reducing and rearrange, it yields:

$$\frac{e^2\alpha^4\pi^4}{8\pi a_e^2\left(1-\tilde{m}_e/\tilde{m}_p\right)^3\tilde{\alpha}^3} \times \frac{1}{T_\circ} = \frac{m_e v_e^2\alpha^2\eta^2\pi^4}{12\left(1-\tilde{m}_e/\tilde{m}_p\right)^{3/2}\tilde{\alpha}} \times \frac{1}{T_\circ}$$

Multiply both flanks by the integer 2, and rearrange, yields:

$$\frac{e^2}{4a_e^2}\left(\frac{\alpha\pi}{\left(1-\tilde{m}_e/\tilde{m}_p\right)}\right)^3 \times \frac{\alpha}{\tilde{\alpha}^2} \times \frac{1}{T_\circ} = \frac{m_e v_e^2\alpha^2\eta^2\pi^4}{6\left(1-\tilde{m}_e/\tilde{m}_p\right)^{3/2}} \times \frac{1}{T_\circ}$$

The terms at each flank represents the Boltzmann constant k .We will use the term at the left flank:

$$(200)\qquad k = \frac{e^2}{4a_e^2}\left(\frac{\alpha\pi}{\left(1-\tilde{m}_e/\tilde{m}_p\right)}\right)^3 \times \frac{\alpha}{\tilde{\alpha}^2} \times \frac{1}{T_\circ}$$

Substitute the following data values into Eq. 200:

Theoretical elementary charge: $e = 1.602188\times10^{-19}[C]$

Initial value of the fine structure constant: $\tilde{\alpha} = 7.315334297\times10^{-3}$

Theoretical value of the Bohr radius: $a_e = 5.291793809\times10^{-11}[m]$

Theoretical value of the fine structure constant: $\alpha = 7.29735579\times10^{-3}$

Theoretical electron reduced mass factor: $\left(1-\tilde{m}_e/\tilde{m}_p\right) = 0.999455756$

Theoretical value of the absolute temperature $T_\circ$ in Kelvin scale: $T_\circ = 273.166[K^\circ]$

Yields the theoretical value of the Boltzmann constant k

$k = 1.380660832\times10^{-23}[J/K^\circ]$

For comparison, the experimentally received value [2] is

$k = 1.38062\times10^{-23}[J/K^\circ]$

7-3. The Molar Gas constant R

Starting with Eq. 200:

$$k = \frac{e^2}{4a_e^2}\left(\frac{\alpha\pi}{\left(1-\tilde{m}_e/\tilde{m}_p\right)}\right)^3 \times \frac{\alpha}{\tilde{\alpha}^2} \times \frac{1}{T_\circ}$$

Substitute $\left[\frac{1}{T_\circ} = \frac{\tilde{\alpha}}{2\sqrt{1-\tilde{\alpha}^2/\beta^2}}\right]$ from Eq. 199 in, forms after reducing:

$$k = \frac{e^2}{8a_e^2}\left(\frac{\alpha\pi}{\left(1-\tilde{m}_e/\tilde{m}_p\right)}\right)^3 \times \frac{\alpha}{\tilde{\alpha}} \times \frac{1}{\sqrt{1-\tilde{\alpha}^2/\beta^2}}$$

Now multiplying the right flank by $\frac{\tilde{m}_p}{\tilde{m}_p}$ yields after rearranging:

$$k = \frac{e^2}{8\tilde{m}_p a_e^2}\left(\frac{\alpha\pi}{\left(1-\tilde{m}_e/\tilde{m}_p\right)}\right)^3 \times \left[\frac{\alpha}{\tilde{\alpha}} \times \frac{\tilde{m}_p}{\sqrt{1-\tilde{\alpha}^2/\beta^2}}\right]$$

The term in the square brackets at the right flank of last equation is the inverse of the Avogadro constant as it was defined at Eq. 194:

(201) $$k = \frac{e^2}{\tilde{m}_p a_e^2}\left(\frac{\alpha\pi}{2\times\left(1-\tilde{m}_e/\tilde{m}_p\right)}\right)^3 \times \frac{1}{N_A}$$

Multiplying both flanks by N_A, forms:

$$k \times N_A = \frac{e^2}{\tilde{m}_p a_e^2}\left(\frac{\alpha\pi}{2\times\left(1-\tilde{m}_e/\tilde{m}_p\right)}\right)^3$$

The left flank is the definition of the Molar Gases constant $R = k \times N_A$

(202) $$R = \frac{e^2}{\tilde{m}_p a_e^2}\left(\frac{\alpha\pi}{2\times\left(1-\tilde{m}_e/\tilde{m}_p\right)}\right)^3$$

Substituting the following data values into Eq. 202
Theoretical elementary charge: $e = 1.602188\times10^{-19}[C]$
Theoretical value of the Bohr radius: $a_e = 5.291793809\times10^{-11}[m]$
Initial proton mass: $\tilde{m}_p = 1.663190691\times10^{-27}[kg]$
Theoretical value of the fine structure constant: $\alpha = 7.29735579\times10^{-3}$
Theoretical electron reduced mass factor: $\left(1-\tilde{m}_e/\tilde{m}_p\right) = 0.999455756$
Yields the theoretical value of the molar gases constant R
$R = 8314.663238[J/kmol\mathrm{K}^\circ] = 8.314663238\times10^3[J/mol\ \mathrm{K}^\circ]$.
For comparison, the experimentally received value [8] is
$R = 8314.5107[J/kmol\mathrm{K}^\circ]$.

8
The sun mass M_S and Kepler constant K

Equating Eq. 52

$$\tilde{G} = 4\pi\,\tilde{\varepsilon}_{\circ}\tilde{\alpha}^2 c^4 \tilde{r}_p^2\left(1-\beta^2\right)$$

and Eq. 45

$$\tilde{G} = \frac{\tilde{e}^2}{4\pi\,\tilde{\varepsilon}_{\circ}\tilde{m}_{pl}^2}$$

yields

(203) $$4\pi\,\tilde{\varepsilon}_{\circ}\tilde{\alpha}^2 c^4 \tilde{r}_p^2\left(1-\beta^2\right) = \frac{\tilde{e}^2}{4\pi\,\tilde{\varepsilon}_{\circ}\tilde{m}_{pl}^2}$$

Multiplying the numerator and the denominator of Eq. 203 by $\frac{1}{\tilde{e}^2}$, along with substituting the relation $\sqrt{1-\beta^2} = \pi^3\beta^2$ from Eq. 38 instead of one of the $\left(1-\beta^2\right)$ components on the left side, yields

(204) $$4\pi\,\tilde{\varepsilon}_{\circ}\tilde{\alpha}^2 c^4 \tilde{r}_p^2\sqrt{1-\beta^2}\left(\pi^3\beta^2\right) = \frac{\tilde{e}^2\left(\frac{1}{\tilde{e}^2}\right)}{4\pi\,\tilde{\varepsilon}_{\circ}\left(\frac{1}{\tilde{e}^2}\right)\tilde{m}_{pl}^2}$$

Placing $\tilde{m}_{pl} = \left(\frac{1}{\tilde{e}}\tilde{m}_p\beta\right)$ from Eq. 41 into the denominator on the right side of Eq. 204, instead of **one** of the $\tilde{m}_{pl}^2$ components, yields

(205) $$4\pi\,\tilde{\varepsilon}_{\circ}\tilde{\alpha}^2 c^4 \tilde{r}_p^2\sqrt{1-\beta^2}\left(\pi^3\beta^2\right) = \frac{1}{4\pi\,\tilde{\varepsilon}_{\circ}\left(\frac{1}{\tilde{e}^2}\right)\tilde{m}_{pl}\left(\frac{1}{\tilde{e}}\tilde{m}_p\beta\right)}$$

Multiplying both sides of Eq. 205 by **2**, along with substituting the relation $c^2\beta^2 = v_p^2$ from Eq. 7 and rearranging, yields

(206) $$8\pi^4\tilde{\varepsilon}_{\circ}\tilde{\alpha}^2 c^2 v_p^2 \tilde{r}_p^2\sqrt{1-\beta^2} = \frac{1}{2\pi\,\tilde{\varepsilon}_{\circ}\left(\frac{1}{\tilde{e}^2}\right)\tilde{m}_{pl}\left(\frac{1}{\tilde{e}}\tilde{m}_p\beta\right)}$$

Substituting the relation $\pi^4 = \frac{4\beta}{\tilde{\alpha}}$ from Eq. 36 into the left side of Eq. 206 and rearranging yields

$$8 \times \frac{4\beta}{\tilde{\alpha}} \tilde{\varepsilon}_\circ \tilde{\alpha}^2 c^2 v_p^2 \tilde{r}_p^2 \sqrt{1-\beta^2} = \frac{1}{2\pi\,\tilde{\varepsilon}_\circ \left(\frac{1}{\tilde{e}^2}\right) \tilde{m}_{pl} \left(\frac{1}{\tilde{e}} \tilde{m}_p \beta\right)}$$

Or, after transposing and rearranging,

$$(207) \qquad 32\tilde{\varepsilon}_\circ^2 \tilde{\alpha}^2 c^2 \tilde{r}_p \tilde{m}_p \left(\frac{1}{\tilde{e}} v_p^2 \tilde{r}_p \sqrt{1-\beta^2}\right) = \left[\frac{\tilde{\alpha}}{2\pi\,\beta^2 \left(\frac{1}{\tilde{e}^2}\right) \tilde{m}_{pl}}\right]$$

Now, substituting Eq. 17 with the *initial components*, along with the relation $c^2\beta^2 = v_p^2$ from Eq. 7, here

$$c\,\tilde{r}_p \tilde{m}_p = \left(\frac{\tilde{h}}{2\pi}\right) \frac{1}{\beta\sqrt{1-\beta^2}}$$

into the left side of Eq. 207 yields

$$(208) \qquad 32\tilde{\varepsilon}_\circ^2 \tilde{\alpha}^2 c \left(\frac{\tilde{h}}{2\pi}\right) \frac{1}{\beta\sqrt{1-\beta^2}} \left(\frac{1}{\tilde{e}} v_p^2 \tilde{r}_p \sqrt{1-\beta^2}\right) = \left[\frac{\tilde{\alpha}}{2\pi\,\beta^2 \left(\frac{1}{\tilde{e}^2}\right) \tilde{m}_{pl}}\right]$$

Substituting Eq. 17a in its *initial components* form into Eq. 16 in its initial form
$\tilde{e}^2 = \tilde{m}_e \tilde{v}_e^2 4\pi \tilde{a}_e \tilde{\varepsilon}_\circ$
along with substituting the relation $\tilde{v}_e = c\tilde{\alpha}$ from Eq. 6 yields

$$(209) \qquad \tilde{h} = \frac{\tilde{e}^2}{2\tilde{\varepsilon}_\circ \tilde{\alpha} c}$$

Substituting Eq. 209 into the left side of Eq. 208 and reducing as needed yields

$$(210) \qquad \frac{16\tilde{\varepsilon}_\circ \tilde{\alpha}\,\tilde{e}^2}{\beta\sqrt{1-\beta^2}} \left(\frac{\frac{1}{\tilde{e}} v_p^2 \tilde{r}_p \sqrt{1-\beta^2}}{2\pi}\right) = \left[\frac{1}{\frac{2\pi\,\beta^2}{\tilde{\alpha}} \left(\frac{1}{\tilde{e}^2}\right) \tilde{m}_{pl}}\right]$$

By choosing components that match the numerical values and the right units, it turns out that the expression in the parenthesis on the left side of Eq. 210 could be an appropriate candidate for the inverse of the *initial Kepler constant* $\tilde{K}$ (the expression $\frac{1}{\tilde{e}}$ is used as a dimensionless counting tool)

$$(211) \qquad \tilde{K} = \frac{2\pi}{\left(\frac{1}{\tilde{e}}\right) v_p^2 \tilde{r}_p \sqrt{1-\beta^2}}$$

By choosing components that match the value and the right units, it turns out that the expression on the right side of Eq. 210 could be an appropriate candidate for the inverse of the *initial sun mass* $\tilde{M}_S$

(the expression $\frac{1}{\tilde{e}^2}$ express an enormous quantity of *Planck mass* which forms the *sun mass*)

$$(212) \qquad \tilde{M}_S = \frac{2\pi\ \beta^2}{\tilde{\alpha}} \left(\frac{1}{\tilde{e}^2}\right) \tilde{m}_{pl}$$

These specific choices of components will be explained by the following developments.
Substituting the initial Kepler constant from Eq. 211 and the initial sun mass from Eq. 212 into Eq. 210 yields

$$\left[\frac{1}{\tilde{K}}\right] \frac{16\tilde{\varepsilon}_o \tilde{\alpha} \tilde{e}^2}{\beta\sqrt{1-\beta^2}} = \left[\frac{1}{\tilde{M}_S}\right]$$

Or, after transposing components,

$$(213) \qquad \left[\frac{\tilde{M}_S}{\tilde{K}}\right] = \frac{\beta\sqrt{1-\beta^2}}{16\tilde{\varepsilon}_o \tilde{\alpha} \tilde{e}^2}$$

The following equations[2,7] represent Newton's universal laws of gravity combined with the Kepler laws, and describe the force activated by the sun on a solar system star. This force depends on the sun's mass, the star's mass and its radius, and the Kepler and gravitation constants

$$F = \frac{4\pi^2}{K}\left(\frac{m}{r^2}\right)$$

and

$$F = GM_s\left(\frac{m}{r^2}\right)$$

F is the force activated by the sun, M_S is the sun mass, m is the star mass, r is the star radius, G is the gravitation constant, and K is the Kepler constant.
Equating these two equations, yields

$$\frac{4\pi^2}{K} = GM_S$$

Using the last equation, but in its *initial values* of components form, dividing both sides of the equality by $\tilde{G}\tilde{K}$ yields

(214) $$\frac{4\pi^2}{\widetilde{G}\widetilde{K}^2} = \frac{\widetilde{M}_S}{\widetilde{K}}$$

Now the argument is this, **if** the choice of components chosen in Eq. 211 and Eq. 212 **is correct,** then the left side of Eq. 214 should be **equal** to the right side of Eq. 213

(215) $$\frac{4\pi^2}{\widetilde{G}\widetilde{K}^2} = \frac{\beta\sqrt{1-\beta^2}}{16\widetilde{\varepsilon}_\circ \widetilde{\alpha}\widetilde{e}^2}$$

For checking, substitute Eq. 52, here
$\widetilde{G} = 4\pi\ \widetilde{\varepsilon}_\circ \widetilde{\alpha}^2 c^4 \widetilde{r}_p^2 \left(1-\beta^2\right)$,
along with Eq. 211, raised to a power of **2**

$$\widetilde{K}^2 = \frac{4\pi^2}{\left[\left(\frac{1}{\widetilde{e}^2}\right) v_p^4 \widetilde{r}_p^2 \left(1-\beta^2\right)\right]}$$

into Eq. 215, yields

(216) $$\frac{4\pi^2}{\left[4\pi\ \widetilde{\varepsilon}_\circ \widetilde{\alpha}^2 c^4 \widetilde{r}_p^2 \left(1-\beta^2\right)\right] \frac{4\pi^2}{\left[\left(\frac{1}{\widetilde{e}^2}\right) v_p^4 \widetilde{r}_p^2 \left(1-\beta^2\right)\right]}} = \frac{\beta\sqrt{1-\beta^2}}{16\widetilde{\varepsilon}_\circ \widetilde{\alpha}\widetilde{e}^2}$$

After reducing and transposing the components on both sides and rearranging, it yields

(217) $$\frac{\beta^4}{4\pi\ \widetilde{\alpha}} = \frac{\beta\sqrt{1-\beta^2}}{16}$$

Now, substituting the relation $\sqrt{1-\beta^2} = \pi^3\beta^2$ from Eq. 38 into the right side of Eq. 217 yields

(218) $$\frac{\beta^4}{4\pi\ \widetilde{\alpha}} = \frac{\beta\pi^3\beta^2}{16}$$

Transposing terms and reducing as needed yields Eq. 36, which is the basic relation of this theory!

$$\frac{\beta}{\widetilde{\alpha}} = \frac{\pi^4}{4}$$

This implies that the choice made for the components of both K and $\widetilde{M}_S$ in Eq. 211 and Eq. 212, respectively, was correct, and any other combination would not have resulted in the last basic relation!

The *final form* of the *Kepler constant* and *sun mass*, which matches the observed astronomy data, should be identical to the form of Eq. 211 and Eq. 212.

So, for the *Kepler constant* K we write

(219) $$K = \frac{2\pi}{\left(\frac{1}{e}\right) v_p^2 r_p \sqrt{1-\beta^2}}$$

Substituting the following data values into Eq. 219

Theoretical value related to the elementary charge acting as a counting tool: $\left(\frac{1}{e}\right) = 1/1.602188 \times 10^{-19}$

Relativistic proton radius: $r_p' = r_p\sqrt{1-\beta^2} = 1.180555339 \times 10^{-15}[m]$

Theoretical proton spinning velocity: $v_p = 5.34065119 \times 10^{7}[ms^{-1}]$

Yields the theoretical value of the *Kepler constant* K

$$K = 2.9896377 \times 10^{-19}[s^2 m^{-3}]$$

For comparison, calculating the Kepler constant based on known values of the sun mass and gravitation constant, substituted into the following equation

$$\frac{4\pi^2}{K} = GM_S$$

Yields

$$K = 2.97445995 \times 10^{-19}[s^2 m^{-3}]$$

For the sun mass M_S we write

(220) $$M_S = \frac{2\pi\beta^2}{\alpha}\left(\frac{1}{e^2}\right) m_{pl}$$

Substituting the following data values into Eq. 220

Theoretical value related to the elementary charge acting as a counting tool: $\left(\frac{1}{e}\right) = 1/1.602188 \times 10^{-19}$

Theoretical fine structure constant: $\alpha = \frac{1}{137.03593859}$

Theoretical value of Planck mass: $m_{pl} = 1.85978 \times 10^{-9}[kg]$

Key constant: $\beta = 0.178145016$

Yields the theoretical value of the *sun mass* M_S

$$M_S = 1.9796862 \times 10^{30}[kg]$$

For comparison, the estimated *sun mass* value[8] is

$$M_S = 1.9891 \times 10^{30}[kg]$$

9
Black Holes and Singularity

The Bekenstein Hawking formula for Black hole Entropy in additional formalism

Start with Eq. 46 with substitution of $\widetilde{m}_{pl} = \left(\frac{1}{\widetilde{e}}\widetilde{m}_p\beta\right)$ from Eq. 41, written as follows

$$G = \frac{\alpha c^2 r_p\sqrt{1-\beta^2}}{\left(\frac{1}{e}\right)\left(\frac{1}{e}\right)m_p\beta}$$

That is equal to Eq. 47

$$G = \frac{l_{pl}c^2}{m_{pl}}$$

Or, after transposing components,

$$\frac{Gm_{pl}}{c^2 l_{pl}} = 1$$

Where m_{pl} is the value of Planck mass from Eq. 41

$$m_{pl} = \left(\frac{1}{e}m_p\beta\right)$$

And $\widetilde{l}_{pl}$ is the Planck length from Eq. 48

$$l_{pl} = \frac{\alpha r_p\sqrt{1-\beta^2}}{\left(\frac{1}{e}\right)}$$

Now, one of the solutions in the general relativity theory, originating from certain boundary conditions, claims the possibility of a star to become a black hole[5]

$$\frac{2GM}{c^2R} = 1$$

Where G, is the Gravitation constant, c is the speed of light in vacuum, M is the star's mass, and R is its radius.
The condition to the creation of a Black hole comes from the following equation[8]:

$$ds^2 = \frac{dx^2}{1-\frac{2GM}{c^2R}} \tag{221}$$

Where ds the infinitesimal distance the ray is passing, and dx its direction coordinate.

The problem existing in this equation is this: when the expression $\frac{2GM}{c^2R}$ in the denominator of the equation is equal to **1**, it implies that the distance that the ray of light may travel is infinite along dx, or, alternatively, a ray of light might take an infinite time to travel this distance. I believe that the

solution for solving this issue should be based on the assumption that the above equation is applicable at Planck dimensions as well.
Assuming that the star is shrinking to a radius that is double Planck length (the integer **2** appears in the original general relativity theory, and is related in my opinion to the relativistic effects related to the macro magnitudes, though in the atomic domain it should be equal to 1, and if so it could be more logical to assume that the radius will be **one Planck length** and not double if **2** is eliminated).
So:

$$R = 2l_{pl}$$

and its mass is equal to one Planck mass

$$M = m_{pl}$$

In these dimensions, the speed of light in the atomic domain c_n is significant, and should replace the component of the speed of light in vacuum c in the condition equation, along with the above definitions for R and M, as follows

$$ds^2 = \frac{dx^2}{1 - \dfrac{2Gm_{pl}}{c_n^2 (2l_{pl})}}$$

Substituting the gravitation constant from Eq. 47

$$G = \frac{l_{pl} c^2}{m_{pl}}$$

into Eq. 221 yields

$$ds^2 = \frac{dx^2}{1 - \dfrac{2l_{pl} c^2 m_{pl}}{c_n^2 m_{pl} (2l_{pl})}}$$

Or after reduction:

$$ds^2 = \frac{dx^2}{1 - \dfrac{c^2}{c_n^2}} \tag{222}$$

It is possible to conclude from Eq. 222 that the star can shrink to Planck dimensions, but it will never reach singularity, since the expression in the denominator is a constant value **smaller than 1**

$$\frac{c^2}{c_n^2} = 0.9986622$$

Therefore eliminating the possibility of ever becoming zero!

Example: An accelerated proton up to $0.99933c_n$ (relatively to the atomic environment, actually represents the speed of light in vacuum c) emits a photon with a velocity of c_n (relatively to the proton). What is the velocity of the photon relatively to the atomic environment?

The data is:

$V_{proton} = 0.99933c_n$ (The velocity of the proton relatively to nucleus environment, which actually is the speed of light in vacuum c)

$\dot{V}_{photon} = c_n$ (The velocity of the photon relatively to proton)

$\ddot{V} = ?$ (The velocity of the photon relatively to atomic environment)

The solution:

(223) $$\ddot{V}_{\substack{photon\ relatively\ to\\ atomic\ environment}} = \frac{\dot{V}_{photon} + V_{proton}}{1 - \dfrac{V_{proton} \times \dot{V}_{photon}}{c_n^2}} = \frac{c_n + 0.99933c_n}{1 + \dfrac{0.99933c_n \times c_n}{c_n^2}} = c_n$$

a. Bekenstein Hawking Entropy of a Black hole at the Planck scale: $S_{BH\,(pl)}$

The general Hawking entropy for a typical black hole is as following:

$$S_{BH} = \frac{A\,k\,c^3}{4\hbar\,G}$$

Substitute: $\hbar = m_p c\beta\, r_p\sqrt{1-\beta^2}$ from Eq. 17b, $G = \dfrac{\alpha\, e\, r_p\sqrt{1-\beta^2}\, c^2}{m_{pl}}$ From Eq. 46 and

$A = 4\pi\, l_{pl}^2$ area of the event horizon: Where l_{pl} is the Planck length: $l_{pl} = \alpha\, e\, r_p\sqrt{1-\beta^2}$ from Eq. 48, it yields:

$$S_{BH\,(pl)} = \frac{4\pi\, l_{pl}^2\, k\, c^3}{4[m_p c\beta\, r_p\sqrt{1-\beta^2}][\alpha\, e\, r_p\sqrt{1-\beta^2}\, c^2/m_{pl}]}$$ Or after reducing

$$S_{BH\,(pl)} = \frac{\pi\, l_{pl}\, k\, m_{pl}}{m_p\beta\, r_p\sqrt{1-\beta^2}}$$

Now multiplying the last equation numerator and denominator by $(\alpha\, e)$ with a slight modification:

$$S_{BH\,(pl)} = \frac{\pi\, l_{pl}\, k\, m_{pl}\,\alpha}{\underbrace{[(1/e)\, m_p\beta]}_{m_{pl}}\underbrace{[\alpha\, e\, r_p\sqrt{1-\beta^2}]}_{l_{pl}}}$$

After reducing it yields the Hawking Entropy of a Black hole at the Planck scale: $S_{BH\,(pl)}$

(224) $$S_{BH\,(pl)} = \pi\,\alpha\, k$$

Note: The presence of the fine structure constant α in the Hawking entropy equation, defines the black hole electrodynamics characteristics!

b. Hawking Radiation Temperature of a Black hole at the Planck scale: $T_{H\,(pl)}$

The general Hawking entropy of a black hole is as following:

$$T_H = \frac{\hbar\, c^3}{8\pi\, k\, G\, M}$$

Substitute the following expressions into the last equation:

Reduced Planck constant $\hbar = m_p v_p r_p\sqrt{1-\beta^2}$ or with $v_p = c\beta \Rightarrow \hbar = m_p c\beta\, r_p\sqrt{1-\beta^2}$, the Planck's mass $M = m_{pl}$ and the gravitational constant G from Eq. 46, it yields:

$$T_{H\,(pl)} = \frac{m_p c\beta\, r_p\sqrt{1-\beta^2}\, c^3\, m_{pl}}{8\pi\, k\,\alpha\, e\, r_p\sqrt{1-\beta^2}\, c^2\, m_{pl}}$$

Or after reducing it yields:

$$\mathrm{T}_{\mathrm{H}(pl)} = (1/e)\, m_p \beta \left(\frac{c^2}{8\pi\, \alpha\, k} \right)$$

And with substitution of $(1/e) m_p \beta = m_{pl}$ from Eq. 41 it yields: $\mathrm{T}_{\mathrm{H}(pl)} = \dfrac{m_{pl}\, c^2}{8\pi\, \alpha\, k}$

Substitute the Bekenstein Hawking entropy $S_{BH\,(pl)} = \pi\, \alpha\, k$ from Eq. 224 yields the Hawking temperature at the Planck scale $\mathrm{T}_{\mathrm{H}(pl)}$:

(225) $$\mathrm{T}_{\mathrm{H}(pl)} = \frac{1}{8} \left(\frac{m_{pl}\, c^2}{S_{BH\,(pl)}} \right)$$

c. The Bekenstein inequality of the Bound State of a Black hole: S_{B}

The general form of the Bekenstein inequality of the Bound state of a Black hole is as following:

$$S_{\mathrm{B}} \leq \frac{2\pi k R\, E}{\hbar\, c}$$

Where S_{B} is the black hole entropy, k is Boltzmann constant, R is the radius of the black hole event horizon, E is energy within the black hole enclosed sphere, $\hbar$ is the reduced Planck constant and c is the speed of light.

d. Bekenstein Bound State of a Black hole at the Planck scale in information terms: $I_{\mathrm{B}(pl)}$

The general Bekenstein bound state of a Black hole in information terms is as following:

$$I_{\mathrm{B}} \leq \frac{2\pi k R\, E}{\hbar\, c \ln 2}$$

Where I_{B} is the information given in the number of bits of the quantum states within the Black Hole, The factor $\ln 2$ converts the result to a logarithm form.

For information in the Plank scale $I_{\mathrm{B}(pl)}$, substitute $R = l_{pl}$, $E = m_{pl}\, c^2$ and $\hbar = m_p\, c\, \beta\, r_p \sqrt{1 - \beta^2}$

$$I_{\mathrm{B}(pl)} \leq \frac{2\pi\, k\, l_{pl}\, m_{pl}\, c^2}{m_p\, c^2 \beta\, r_p \sqrt{1 - \beta^2}\, \ln 2}$$

Now multiplying the last equation numerator and denominator by $(\alpha\, e)$ with a slight modification:

$$I_{\mathrm{B}(pl)} \leq \frac{2\pi\, k\, l_{pl}\, m_{pl}\, c^2\, \alpha}{\underbrace{[(1/e) m_p \beta]}_{m_{pl}}\, c^2\, \underbrace{[\alpha\, e\, r_p \sqrt{1 - \beta^2}]}_{l_{pl}}\, \ln 2} \;\rightarrow\; I_{\mathrm{B}(pl)} \leq \frac{2\pi\, \alpha\, k}{\ln 2}$$

Or substitute the Eq.224 for the Bekenstein Hawking Entropy at the Planck scale $S_{BH\,(pl)} = \pi\, \alpha\, k$ yields:

(226) $$I_{\mathrm{B}(pl)} \leq \frac{2}{\ln 2} \times S_{BH\,(pl)}$$

e. Bekenstein Hawking Entropy for a typical Black hole: S_{BH}

The general Hawking entropy of a black hole is as following:

$$S_{BH} = \frac{\mathrm{A}\, k\, c^3}{4\, \hbar\, G}$$

Substitute $\hbar$ and G in, yields:

$$S_{BH} = \frac{\mathrm{A}\,k\,c^3}{4[m_p c\beta\, r_p\sqrt{1-\beta^2}][\alpha\, e\, r_p\sqrt{1-\beta^2}c^2\,(1/m_{pl})]}$$

Multiplying the last equation numerator and denominator by $(\pi\,\alpha\,e)$ with a slight modification, and reducing the speed of light c and the Planck mass m_{pl} yields:

$$S_{BH} = \frac{\mathrm{A}\,\cancel{c}^3\,\cancel{m}_{pl}\,\pi\,\alpha\,k}{4\pi\,\underbrace{[(1/e)m_p\beta]}_{m_{pl}}\,\underbrace{[\alpha\, e\, r_p\sqrt{1-\beta^2}]}_{l_{pl}}\,\cancel{c}^3\,\underbrace{[\alpha\, e\, r_p\sqrt{1-\beta^2}]}_{l_{pl}}} \rightarrow S_{BH} = \frac{\mathrm{A}\,\pi\,\alpha\,k}{4\pi\,l_{pl}^2}$$

Substitute Eq. 224 result for the Bekenstein Hawking entropy at the Planck scale $S_{BH\,(pl)} = \pi\,\alpha\,k$ yields the Bekenstein Hawking Entropy for a typical Black hole:

(227) $$S_{BH} = \frac{\mathrm{A}}{4\pi\,l_{pl}^2} \times S_{BH\,(pl)}$$

Note: A is the typical area of a black hole event horizon, and $4\pi\,l_{pl}^2$ is the area of a black hole event horizon at the Planck scale!

f. Hawking Radiation Temperature for a typical Black hole: $\mathrm{T_H}$

The general Hawking temperature of a black hole is as following:

$$\mathrm{T_H} = \frac{\hbar\,c^3}{8\pi\,k\,G\,M}$$

Substitute $\hbar$ and G in, yields:

$$\mathrm{T_H} = \frac{m_p c\beta\, r_p\sqrt{1-\beta^2}\,c^3}{8\pi\,k[\alpha\, e\, r_p\sqrt{1-\beta^2}c^2\,(1/m_{pl})]\ \mathrm{M}}$$

Multiplying the last equation numerator and denominator by $(\alpha\,e)$ with a slight modification, and reducing the speed of light c and the Planck radius l_{pl} yields:

$$\mathrm{T_H} = \frac{\overbrace{[(1/e)\,m_p\beta]}^{m_{pl}}\overbrace{[\alpha\, e\, r_p\sqrt{1-\beta^2}\,]}^{l_{pl}}m_{pl}\,c^2}{8\pi\,\alpha k\,\underbrace{[\alpha\, e\, r_p\sqrt{1-\beta^2}\ ]}_{l_{pl}}\ \mathrm{M}} \rightarrow \mathrm{T_H} = \frac{m_{pl}^2\,c^2}{8\pi\,\alpha\,k\,\mathrm{M}}$$

Reduce l_{pl} and substitute Eq. 224 result for the Bekenstein Hawking entropy at the Planck scale $S_{BH\,(pl)} = \pi\,\alpha\,k$ in with a slight modification, yields:

$$\mathrm{T_H} = \frac{m_{pl}}{\mathrm{M}}\left(\frac{m_{pl}\,c^2}{8S_{BH\,(pl)}}\right)$$

Substitute Eq. 225 result for the Hawking radiation temperature at the Planck scale $\left(\frac{m_{pl}\,c^2}{8S_{BH\,(pl)}}\right) = \mathrm{T}_{\mathrm{H}(pl)}$, yields the Hawking radiation temperature for a typical Black hole:

(228) $$\mathrm{T_H} = \frac{m_{pl}}{\mathrm{M}} \times \mathrm{T}_{\mathrm{H}(pl)}$$

g. Bekenstein Bound state of a typical Black hole in information terms: I_{B}

The general Bekenstein bound state of a Black hole in information terms is as following:

$$I_{\mathrm{B}} \leq \frac{2\pi\, k\, R\, E}{\hbar\, c \ln 2}$$

Substitute $\hbar$ in, yields:

$$I_{\mathrm{B}} \leq \frac{2\pi\, k\, R\, E}{[m_p c \beta\, r_p \sqrt{1 - \beta^2}\,]\, c \ln 2}$$

Multiplying the last equation numerator and denominator by $(\alpha\, e)$ with a slight modification, yields:

$$I_{\mathrm{B}} \leq \frac{2\pi\, \alpha\, k\, R\, E}{\underbrace{[(1/e) m_p \beta]}_{m_{pl}}\, c^2\, \underbrace{[\alpha\, e\, r_p \sqrt{1 - \beta^2}\,]}_{l_{pl}} \ln 2} \rightarrow I_{\mathrm{B}} \leq \frac{2\pi\, \alpha\, k\, R\, E}{m_{pl}\, c^2\, l_{pl} \ln 2}$$

Substitute Eq. 224 result for the Bekenstein Hawking entropy at the Planck scale $S_{BH(pl)} = \pi\, \alpha\, k$ in, yields:

$I_{\mathrm{B}} \leq \dfrac{2 S_{BH\,(pl)}\, R\, E}{m_{pl}\, c^2\, l_{pl} \ln 2}$ Or with substituting $\mathrm{T}_{\mathrm{H}(pl)} = \dfrac{1}{8}\left(\dfrac{m_{pl}\, c^2}{S_{BH\,(pl)}}\right)$ from Eq. 225 in, yields:

(229) $$I_{\mathrm{B}} \leq \frac{1}{4 \times \ln 2} \times \frac{R}{l_{pl}} \times \frac{E}{\mathrm{T}_{\mathrm{H}\,(pl)}}$$

Now equating Eq. 222 and Eq. 221, yields:

$$1 - \frac{c^2}{c_n^2} = 1 - \frac{2GM}{c^2 R} \rightarrow \frac{c^2}{c_n^2} = \frac{2GM}{c^2 R}$$

Substitute the gravitational constant G from Eq. 46, yields:

$$\frac{c^2}{c_n^2} = \frac{2[\alpha\, e\, r_p \sqrt{1 - \beta^2}\,]\cancel{c}^2 M}{m_{pl}\, \cancel{c}^2 R} \rightarrow \frac{c^2}{c_n^2} = \frac{2 l_{pl} M}{R\, m_{pl}} \rightarrow \text{Black hole mass: } M = \frac{R \times m_{pl} \times c^2}{2 l_{pl} \times c_n^2}$$

Note: c_n was found in Eq. 122 as the speed of light in the atomic domain!

The Eq. 228 result of the Hawking temperature for a typical Black hole depended on its mass M is:

$$\mathrm{T}_{\mathrm{H}} = \frac{m_{pl}}{M} \times \mathrm{T}_{\mathrm{H}\,(pl)}$$

Substitute the previous result for M in, yields the Hawking temperature for a typical Black hole that is depended on the black hole radius R:

$$\mathrm{T}_{\mathrm{H}} = \frac{2 \cancel{m}_{pl} l_{pl} c_n^2}{R\, \cancel{m}_{pl} c^2} \times \mathrm{T}_{\mathrm{H}\,(pl)}$$

(230) $$\mathrm{T}_{\mathrm{H}} = \frac{2 l_{pl}}{R} \times \frac{c_n^2}{c^2} \times \mathrm{T}_{\mathrm{H}\,(pl)}$$

Substitute the Eq. 225 result for the Hawking temperature at the Planck scale $\mathrm{T}_{\mathrm{H}\,(pl)} = \dfrac{1}{8}\left(\dfrac{m_{pl}\, c^2}{S_{BH\,(pl)}}\right)$ in, yields:

$$\mathrm{T}_{\mathrm{H}} = \frac{2 l_{pl} c_n^2}{R\, \cancel{c}^2} \times \frac{m_{pl}\, \cancel{c}^2}{8 S_{BH\,(pl)}} \rightarrow \mathrm{T}_{\mathrm{H}} = \frac{l_{pl} m_{pl} c_n^2}{4 R S_{BH\,(pl)}} \text{ Or} \rightarrow S_{BH\,(pl)} = \frac{l_{pl} m_{pl} c_n^2}{4 R \mathrm{T}_{\mathrm{H}}}$$

The previous result of the Bekenstein Hawking Entropy for a typical Black hole S_{BH} found in Eq. 227 is:

$$S_{BH} = \frac{A}{4\pi l_{pl}^2} \times S_{BH(pl)}$$

The A stands for the typical area of a Black hole event horizon, and it equals to: $A = 4\pi R^2$ Substitute the Bekenstein Hawking entropy at the Planck scale $S_{BH(pl)}$ from Eq. 224 along with $A = 4\pi R^2$ in, yields:

$$S_{BH} = \frac{4\pi R^2}{4\pi l_{pl}^2} \times \frac{l_{pl} m_{pl} c_n^2}{4R\mathrm{T_H}}$$

Or after reducing and a slight modification, the Bekenstein Hawking entropy S_{BH} for a typical black hole is depended on its radius R and the Hawking temperature $\mathrm{T_H}$:

(231) $$S_{BH} = \frac{R}{\mathrm{T_H}}\left(\frac{m_{pl} c_n^2}{4 l_{pl}}\right)$$

10
The Unification of the Gravitational Force and the Electrical Force at Planck Dimensions

Assuming that two entities have Planck mass each and are at Planck length apart, and the attraction force between them is according to the universal gravity law,

(232) $$F = \frac{Gm_{pl}^2}{l_{pl}^2}$$

and the repulsion force between them is according to Coulomb law (where the electrical charge is related to the Planck mass, defined as a collection of the positive protons),

(233) $$F = \frac{e^2}{4\pi\varepsilon_\circ l_{pl}^2}$$

then equating both forces yields

(234) $$\frac{Gm_{pl}^2}{l_{pl}^2} = \frac{e^2}{4\pi\varepsilon_\circ l_{pl}^2}$$

Substituting in Eq. 228 the gravitation constant expression from Eq. 47

$$G = \frac{l_{pl}c^2}{m_{pl}}$$

and reducing the length and the Planck mass components at both sides after that, yields

(235) $$l_{pl}c^2 m_{pl} = \frac{e^2}{4\pi\ \varepsilon_\circ}$$

Substituting one of the m_{pl} expression from Eq. 41

$$m_{pl} = \left(\frac{1}{e} m_p \beta\right)$$

along with substituting the Planck length l_{pl} expression from Eq. 48

$$l_{pl} = \frac{\alpha r_p \sqrt{1-\beta^2}}{\left(\frac{1}{e}\right)}$$

on the left side of Eq. 229, and reducing as needed, yields

(236) $$\frac{\alpha r_p\sqrt{1-\beta^2}}{\left(\frac{1}{e}\right)}\left(\frac{1}{e}m_p\beta\right)c^2 = \frac{e^2}{4\pi\ \varepsilon_\circ}$$

Reducing the $1/e$ component on the left side of Eq. 227 yields after transposing components the expression of Eq. 22

$$e^2 = m_p c^2 \beta\alpha 4\pi\ \varepsilon_\circ r_p\sqrt{1-\beta^2}$$

This means that **at Planck dimensions the gravitational force equals the electrical force**!

11
Schwarzschild Radius

As seen earlier, the condition for the creation of a black hole is[5]

$$\frac{2GM}{c^2R}=1$$

originating from the equation

$$ds^2=\frac{dx^2}{1-\frac{2GM}{c^2R}}$$

There are two important notes for further development:

a. The radius R in the creation of a black hole equation is commonly named the **Schwarzschild radius**[5], after the German scientist who found the conditions to its creation while trying to solve the equations of the general relativity theory for certain conditions. This radius is therefore marked as $R_{Schwarzschild}$, and it is the radius of a star defined as a black hole, constituting a horizon beyond which no escape of any kind of material or energy is possible.

b. As stated earlier, the speed of light in the atomic domain c_n has a significant impact at Planck dimensions, therefore replacing the component of the speed of light in vacuum c in the equations that follows.

Following the above, the equation of the condition for the creation of a black hole receives this form

$$ds^2=\frac{dx^2}{\left(1-\frac{2GM}{c_n{}^2R_{Schwarzschild}}\right)} \tag{237}$$

Substituting the *Schwarzschild radius*, presented here for convenient reasons as $R_{Schwarzschild}=2r$, the star mass $M=m_{star}$, and the gravitation constant expression from Eq. 47

$$G=\frac{l_{pl}c^2}{m_{pl}}$$

into Eq. 228 yields

$$ds^2=\frac{dx^2}{\left[1-\frac{2\left(\frac{l_{pl}c^2}{m_{pl}}\right)m_{star}}{c_n^2(2r)}\right]}$$

or, after reducing and rearranging,

(238) $$ds^2 = \frac{dx^2}{1 - \frac{c^2}{c_n^2}\left[\left(\frac{l_{pl}}{r}\right)\left(\frac{m_{star}}{m_{pl}}\right)\right]}$$

In order to stay consistent with the results of Eq. 222, that is eliminating the possibility of singularity, the expression in the parenthesis in the right term of Eq. 238 must be equal to **1**

$$\left(\frac{l_{pl}}{r}\right)\left(\frac{m_{star}}{m_{pl}}\right) = 1$$

Or, after transposing components

(239) $$r = m_{star}\left(\frac{l_{pl}}{m_{pl}}\right)$$

Substituting the last expression for r into the presentation of the Schwarzschild radius as $R_{schwarzschild} = 2r$, yields the expression for calculating the Schwarzschild radius for any given star, depending only on the star mass, Planck mass, and Planck length

(240) $$R_{schwarzschild} = 2\left[m_{star}\left(\frac{l_{pl}}{m_{pl}}\right)\right]$$

For example, to solve Eq. 231 for the sun

$$R_{schwarzschild} = 2\left(M_S \frac{l_{pl}}{m_{pl}}\right)$$

Substitute the following data values into the last equation

Theoretical sun mass: $M_S = 1.9891 \times 10^{30}\,[kg]$
Theoretical Planck mass: $m_{pl} = 1.85978 \times 10^{-9}\,[kg]$
Theoretical Planck length: $l_{pl} = 1.38027412 \times 10^{-36}\,[m]$

Yielding the theoretical Schwarzschild radius of the Sun
$R_{Schwarzschild\ of\ the\ Sun} = 2938.53[m]$
For comparison, the estimated value[7] is
$R_{Schwarzschild\ of\ the\ Sun} = 2953.49[m]$

12
The Quarks

While developing the expressions for the 'Proton Magnetic Moment to Nuclear Magneton ratio' μ_p/μ_N and the 'Neutron Magnetic Moment to Nuclear Magneton ratio' μ_n/μ_N, I have noticed that these constants reveal relations that might imply on their inner structure. The general magnetic moments expressions can be written as following:

$$\mu_M = \frac{e \times \hbar}{2 \times m} \times (\text{relevant constant}) \quad \text{Or} \quad \mu_M = \frac{e\,v\,r}{2} \times (\text{relevant constant})$$

μ_M -is the magnetic moment of the relevant particle.

e -is the elementary charge of the relevant particle.

m -is the mass of the relevant particle.

$\hbar$ -is the reduced Planck constant $(h/2\pi)$.

v -is the relevant particle's velocity.

r -is the relevant particle's radius.

For the electron:

$$\mu_e = \frac{e \times \hbar}{2 \times m_e} \times \left[\frac{1 - \tilde{m}_e / \tilde{m}_p}{1 - \tilde{\alpha}^2 / \beta^2}\right] \quad \text{where} \quad \left[\frac{1 - \tilde{m}_e / \tilde{m}_p}{1 - \tilde{\alpha}^2 / \beta^2}\right] = 1.00115965$$

For the proton: (Please see Part 4 Chapter 3)

$$\mu_p = \frac{e\,v_p r_p \sqrt{1-\beta^2}}{2} \times \left[\frac{1}{2\beta} \times \frac{\alpha^2}{\tilde{\alpha}^2}\right] \quad \text{Where} \quad \left[\frac{1}{2\beta} \times \frac{\alpha^2}{\tilde{\alpha}^2}\right] = 2.792847386$$

For the neutron: (Please see Part 4 Chapter 4)

$$\mu_n = \frac{e\,v_p r_n \sqrt{1-\beta^2}}{2} \times \left[\frac{1}{3\beta\sqrt{1-\beta^2} \times (1 - \tilde{m}_e/\tilde{m}_p)^2} \times \left(\frac{\tilde{\alpha}^2}{\alpha^2}\right)\right]$$

$$\text{Where} \quad \left[\frac{1}{3\beta\sqrt{1-\beta^2} \times (1 - \tilde{m}_e/\tilde{m}_p)^2} \times \left(\frac{\tilde{\alpha}^2}{\alpha^2}\right)\right] = 1.91304275$$

If we write the last formulas, in slightly modified form, it seems that we can make further inquiries regarding the internal structure of the proton and neutron that these formulas imply: for the proton $\frac{e\,v_p}{2} \times \left[\frac{r_p\sqrt{1-\beta^2}}{2\beta}\right] \times \frac{\alpha^2}{\tilde{\alpha}^2}$ and for the neutron: $\frac{e\,v_p}{2} \times \left[\frac{r_n\sqrt{1-\beta^2}}{3\beta\sqrt{1-\beta^2}}\right] \times \frac{1}{(1-\tilde{m}_e/\tilde{m}_p)^2} \times \left(\frac{\tilde{\alpha}^2}{\alpha^2}\right)$. In this form the terms in the square brackets at both formulas implies that there are three energy levels or orbitals within the proton and the neutron on which the quarks are orbiting. We can observe that the products 2β and 3β at their denominators implies on the quarks velocities at the second and third energy levels respectively (if each multiplied by the speed of light in vacuum c we receive $2c\beta = 2v_p$ and $3c\beta = 3v_p$ respectively), and the fractions $r_p/2$ and $r_n/3$ implies on the quarks orbitals radius at the second and third energy levels respectively. We can assume from this the existence of a quark's velocity $c\beta = v_p$ and an orbital radius r_p related to the first energy level!

Moreover, because there are three orbitals it seems that each quark share one-third of the proton or the neutron mass apiece!

The quark's charge expression should be presented in the form shown in Eq. 18, but with the quarks' parameters related to its orbital. I am introducing Eq. 18 here:

$$e^2 = \frac{m_p v_p^2 4\pi r_p^2 (1-\beta^2)}{6} \times \left(\frac{m_p}{m_e}\right)$$

And with using Eq. 17b:

$$\hbar = m_p v_p r_p \sqrt{1-\beta^2}$$

We can express the square of the elementary charge (For the proton) as following:

(241) $$e^2 = \frac{m_p v_p^2 4\pi r_p^2 (1-\beta^2)}{6}\left(\frac{m_p}{m_e}\right) = m_p v_p r_p \sqrt{1-\beta^2} \times m_p v_p r_p \sqrt{1-\beta^2} \times \left(\frac{4\pi}{6m_e}\right) = \left[\hbar^2\left(\frac{4\pi}{6m_e}\right)\right]$$

The quarks' Orbital Angular Momentum $\vec{L}_{q(\mathrm{n})}$ is calculated by the following equation which varies per each orbital quantum number **n**!

(242) $$\vec{L}_{q(\mathrm{n})} = \mathrm{n}(m_q \times v_{q(\mathrm{n})} \times r_{\mathrm{n}})$$

Where m_q is the quark's mass, r_{n} the orbital radius and $v_{q(\mathrm{n})}$ the orbital velocity are both depend on the quantum number **n**, where **n** = 1, 2, 3. It was revealed from the magnetic momentums of the proton and the neutron that the quark's velocity is $v_{q(\mathrm{n})} = c(\mathrm{n}\beta)$ and the orbital radius is $r_{\mathrm{n}} = r_p\sqrt{1-\beta^2}/\mathrm{n}$ The expression of the quark's charge q_{n} using m_q, $v_{q(\mathrm{n})}$ and r_{n} that corresponds with Eq. 241 and Eq. 242 is:

(243) $$q_{\mathrm{n}}^2 \propto \frac{m_q v_{q(\mathrm{n})}^2 4\pi r_{\mathrm{n}}^2}{6}\left(\frac{m_q}{m_e}\right) \rightarrow q_{\mathrm{n}}^2 = \mathrm{n}(m_q v_{q(\mathrm{n})} r_{\mathrm{n}}) \times \mathrm{n}(m_q v_{q(\mathrm{n})} r_{\mathrm{n}}) \times \left(\frac{4\pi}{6m_e}\right) = \vec{L}_{q(\mathrm{n})}^2\left(\frac{4\pi}{6m_e}\right)$$

The proton consists of three quarks, hence if assuming that each quark share one-third of the proton mass apiece $m_q = 1/3\, m_p$, then substitute m_q along with $v_{q(\mathrm{n})} = c(\mathrm{n}\beta)$ and $r_{\mathrm{n}} = r_p\sqrt{1-\beta^2}/\mathrm{n}$ in Eq. 242 yields the quark's angular momentum $\vec{L}_{q(\mathrm{n})}$ and by using the result from Eq. 242 in Eq. 243 it yields the charge q_{n} at each orbital level **n**.

a. The **first level n = 1** is the outer orbital that outlines the surface of the proton or the neutron.

Substituting: - $m_q = \frac{1}{3}m_p$, $v_{q(\mathrm{n}=1)} = c\beta = v_p$ and $r_{\mathrm{n}=1} = r_p\sqrt{1-\beta^2}$ in Eq. 242, yields the quark's orbital angular momentum for the first level:

(244) $$\vec{L}_{q(\mathrm{n}=1)} = 1\left[\frac{1}{3}m_p v_p r_p\sqrt{1-\beta^2}\right] = \frac{1}{3}\hbar$$

Substituting Eq. 244 result in Eq. 243 and by using the result of Eq. 241 it yields the quarks' charge at the first level:

$$q_{\mathrm{n}=1}^2 = \vec{L}_{q(\mathrm{n}=1)}^2\left(\frac{4\pi}{6m_e}\right) = \left[\frac{1}{3}\hbar\right]^2\left(\frac{4\pi}{6m_e}\right) = \frac{1}{9}\left[\hbar^2\left(\frac{4\pi}{6m_e}\right)\right] = \frac{1}{9}e^2$$

Or after extracting the square root, we receive:

(245) $$q_{\mathrm{n}=1} = \pm\frac{1}{3}e^{\pm}$$

b. The **second level n = 2** is the middle orbital in the proton or the neutron.

Substituting: - $m_q = \frac{1}{3}m_p$, $v_{q(\mathrm{n}=2)} = 2(c\beta) = 2v_p$ and $r_{\mathrm{n}=2} = r_p\sqrt{1-\beta^2}/2$ in Eq. 242, yields the quark's orbital angular momentum for the second level:

(246) $$\vec{L}_{q(\mathrm{n}=2)} = 2\left[\frac{1}{3}m_p \times 2v_p \times r_p\sqrt{1-\beta^2}/2\right] = \frac{2}{3}\hbar$$

Substituting Eq. 246 result in Eq. 243 and by using the result of Eq. 241 it yields the quark's charge at the second level:

$$q^2_{n=2} = \vec{L}^2_{q(n=2)}\left(\frac{4\pi}{6m_e}\right) = \left[\frac{2}{3}\hbar\right]^2\left(\frac{4\pi}{6m_e}\right) = \frac{4}{9}\left[\hbar^2\left(\frac{4\pi}{6m_e}\right)\right] = \frac{4}{9}e^2$$

Or after extracting the square root we receive:

(247) $$q_{n=2} = \pm\frac{2}{3}e^{\pm}$$

c. The **third level n = 3** is the most inner orbital in the proton or neutron.

Substituting: - $m_q = \frac{1}{3}m_p$, $v_{q(n=3)} = 3(c\beta) = 3v_p$ and $r_{n=3} = r_p\sqrt{1-\beta^2}/3$ in Eq. 242, yields the quark's orbital angular momentum for the third level:

(248) $$\vec{L}_{q(n=3)} = 3\left[\frac{1}{3}m_p \times \cancel{3}v_p \times r_p\sqrt{1-\beta^2}/\cancel{3}\right] = \hbar$$

Substituting Eq. 248 result in Eq. 243 and by using the result of Eq. 241 it yields the quark's charge at the third level:

$$q^2_{n=3} = \vec{L}^2_{q(n=3)}\left(\frac{4\pi}{6m_e}\right) = [\hbar]^2\left(\frac{4\pi}{6m_e}\right) = \left[\hbar^2\left(\frac{4\pi}{6m_e}\right)\right] = e^2$$

Or after extracting the square root we receive:

(249) $$q_{n=3} = \pm e^{\pm}$$

Notes:

1. **The third level plays a major role in the weak force decaying process and the subatomic particles interactions.** The significance of the third energy level will be clarified later in the next chapters on the discussion of the origin of the weak force $W^{\pm}$ boson!
2. **The structure of the proton** according to this model is similar (by content) to quark's theory: Two *up* quarks $\left(+\frac{2}{3}e^{+}\right)$ at the second level and one *down* quark $\left(-\frac{1}{3}e^{-}\right)$ at the first level, that their algebraic sum of charges determines the charge of the proton e^{+}.
3. **The structure of the neutron** according to this model is similar (by content) to the quark's theory: One *up* quark $\left(+\frac{2}{3}e^{+}\right)$ at the second level and two *down* quarks $\left(-\frac{1}{3}e^{-}\right)$ at the first level that their algebraic sum of charges determines the neutron **neutral** charge.
4. **The quark's orbital angular momentum** $\vec{L}_q$ at the first and the second levels of the baryons is a factor of one third and two thirds of $\hbar$ according to Eq. 244 and Eq. 246. The same is for the quark's charge according to Eq. 245 and Eq. 247 is one third and two thirds of the $e^{\pm}$ charge.
5. **The true nature of the electrical charge:** The subatomic particles charges in general and the quark's charge in particular are actually the expression of their orbital angular momentum $\vec{L}_q$ and their positive or negative values depend on their revolving directions (Eq. 241 and Eq. 243 show this relation as following here). The Orbital Angular Momentum $(\pm\hbar)$ and $(\pm\vec{L}_{q(n)})$ are vectors properties, meaning that the signs $(\pm)$ are indicating their revolving directions!

$$e^2 = \hbar^2 \times \frac{4\pi}{6m_e} \rightarrow e = \pm\hbar\times\left(\frac{4\pi}{6m_e}\right)^{1/2} \quad \text{and} \quad q^2_n = \vec{L}^2_{q(n)}\left(\frac{4\pi}{6m_e}\right) \rightarrow q_n = \pm\vec{L}_{q(n)}\left(\frac{4\pi}{6m_e}\right)^{1/2}$$

In baryons and mesons $\vec{L}^*_q$ is the algebraic sum of all the angular momentum of their quarks. If a Total orbital angular momentum $\vec{L}^*_q$ of a particle for instance is $(-\hbar\downarrow)$ it expresses that it has a

e^- Charge, and a total orbital angular momentum $\vec{L}_q^*$ of $(+\hbar\uparrow)$ expresses that it has a e^+ charge. If the orbital angular momentum $\vec{L}_q^*$ sum up to zero (offset), it means that the particle is neutral!

5.1 The j^P spin of a baryon **is not** an indicator of its charge. The neutron has j^P spin $(1/2\hbar\uparrow)$ and zero $\vec{L}_q^*$ so it is neutral.

Note: The leptons have an inner structure that is presented in this theory (see chapter 16) that shows their relation to the Pi mesons which reveals the reason of their intrinsic electric charge!

6. **Why the angular momentum of an electron at the ground state in the Hydrogen atom appears to be zero according to the quantum mechanics** (that leads to a conclusion that the electron doesn't revolve at this orbital!):

Explanation: The total angular momentum within the proton is $(+\hbar\uparrow)$ that expresses its e^+ charge (this is the algebraic sum of two *up* quarks with $(+2/3\hbar\uparrow)$ at its second level and one *down* quark $(-1/3\hbar\downarrow)$ at its first level). The proton and the electron in the Hydrogen atom act as one entity and therefore if the electron revolves at the ground state with an orbital angular momentum $(-\hbar\downarrow)$ that expresses an e^- charge (This e^- charge is an induced effect that the electron creates while it revolves at the ground state with an angular momentum and it is not related to the electrons' intrinsic e^- charge originated from its internal structure explained at chapter 16!) The total orbital angular momentum within the proton $(+\hbar\uparrow)$ and the orbital angular momentum $(-\hbar\downarrow)$ of the electron at the ground state of the hydrogen atom are offset and it reflects outside a zero angular momentum!
In addition there are no projections components of the orbital angular momentum on z direction coordinate because the revolving axis of the electron at the ground state in a magnetic field is perpendicular to the field direction, so the angular momentum is wrongly assumed to be zero.

Another aspect of this issue is a discussion over the total electron energy values E_n (Kinetic and Potential) of the orbitals in the Hydrogen atom that for the ground state is:

$$E_{n=1} = -\frac{m_e e^4}{8\varepsilon_o^2 h^2} = -13.6[eV].$$

* It seems difficult to imagine the electron having a kinetic energy by being stationary at its ground state orbital!

7. **The actual role of the orbital angular momentum in the atom structure** (an insight from the joint quantum mechanics formulation and the findings of this theory): There are four quantum numbers that express the atom structure according to the quantum mechanics. The three that are relevant for our discussion are the following (the forth involves the particle's spin):

The Principle Quantum Number (Shells): n = 1, 2, 3...
The Angular Momentum Quantum Number (Sub shells): $l = 0,1,2,....,(n-1)$
The Magnetic Quantum Number values are: $m_l = -l,-(l-1),...,0,....,+(l-1),+l$
The electron's orbital angular momentum: $\vec{L}_l = \sqrt{l(l+1)} \times \hbar$
The projection components on z coordinate of the orbital angular momentum in a magnetic field $\vec{L}_z = \pm m_l\hbar$ (this value can actually be measured and it is important for our discussion).

Some examples:

a. $n = 1 ; l = 0 : \rightarrow m_{l=0} = 0$

$$\underbrace{\vec{L}_z = 0}_{\text{2 electrons one orbital at subshell s}} \quad ;$$

b. $n = 2 ; l = 0, l = 1 : \rightarrow m_{l=0} = 0; m_{l=1} = -1,0,1$

$$\underbrace{\vec{L}_{l=0} = 0}_{\text{2 electrons in one orbital at subshell s}} \; ; \; \underbrace{\vec{L}_{l=1} = \overbrace{(-\hbar \downarrow)}^{\text{inducing } e^- \text{ charge}}, 0, \overbrace{(+\hbar \uparrow)}^{\text{inducing } e^+ \text{ charge}}}_{\text{6 electrons in three orbital at subshell p}} \; ;$$

c. $n = 3 ; l = 0, l = 1, l = 2 : \rightarrow m_{l=0} = 0; m_{l=1} = -1,0,1; m_{l=2} = -2,-1,0,1,2$

$$\underbrace{\vec{L}_{l=0} = 0}_{\text{2 electrons in one orbital at subshell s}} \; ; \; \underbrace{\vec{L}_{l=1} = \overbrace{(-\hbar \downarrow)}^{\text{inducing } e^- \text{ charge}}, 0, \overbrace{(+\hbar \uparrow)}^{\text{inducing } e^+ \text{ charge}}}_{\text{6 electrons in three orbitals at subshell p}} \; ; \; \underbrace{\vec{L}_{l=2} = \overbrace{(-2\hbar \downarrow)}^{\text{inducing } 2e^- \text{ charge}}, \overbrace{(-\hbar \downarrow)}^{\text{inducing } e^- \text{ charge}}, 0, \overbrace{(+\hbar \uparrow)}^{\text{inducing } e^+ \text{ charge}}, \overbrace{(+2\hbar \uparrow)}^{\text{inducing } 2e^+ \text{ charge}}}_{\text{10 electrons in five orbitals at subshell d}}$$

* **The attraction between two electrons with an opposite orbital angular momentum (revolving in an opposite directions) at a given orbital.**

There is attraction between two electrons in an orbital that is created by the induced charges evolves from their opposite orbital angular momentum $(+\hbar \uparrow)$ and $(-\hbar \downarrow)$ despite of their intrinsic e^- charges which supposedly they are meant to repulse each other. For instance, an electron with orbital angular momentum $(+\hbar \uparrow)$ that induces e^+ charge attracts an electron with an opposite orbital angular momentum $(-\hbar \downarrow)$ that induces e^- charge. This attraction keeps the electrons in the orbital attached to a given subshell (s, p, d, f…) which in turn it keeps the electrons attached to the principal level **n** while electrons at lower **n** orbitals create a shielding around the nucleus which decreases the interaction with it. This attraction also creates the chemical bonds while valence electrons are lacking of their opposite companions in the outer atom subshells and they attract electrons from a different atom with an opposite orbital angular momentum (revolves in an opposite directions) to complete their orbitals by creating a covalence or ionic bonds.

For Example: The **Cooper pair** which consists of two electrons is attracting each other if one of them revolves in a direction with orbital angular momentum $(+\hbar \uparrow)$ that induces e^+ charge and the other revolves in an opposite direction with orbital angular momentum $(-\hbar \downarrow)$ that induces e^- charge. This attraction occurs despite of the fact that the electrons possess identical intrinsic e^- charges originated from their internal structure (see chapter 16) which supposedly they are meant to repulse each other.

1. **In general:** Two subatomic particles in contact with an 'opposite orbital angular momentum' (revolving in opposite directions) attract each other, whereas with identical 'orbital angular momentum' (revolving in the same direction) repulse each other!

8. **The true nature of the electron mass (and actually the meaning of the mass in general):**

The mass of the electron can be derived from Eq. 241 as a function of the square of the internal orbital angular momentum $\hbar$ subtracted by the square of its intrinsic electric charge e^- originated from its internal structure (explained at chapter 16) as following:

$$e^2 = \hbar^2 \times \frac{4\pi}{6m_e} \rightarrow m_e = \left(\frac{\hbar}{e}\right)^2 \times \left(\frac{2\pi}{3}\right)$$

Note: The values of the electron mass, Planck constant and the elementary electric charge that is presented in the last equation is in fact **their initial values**. The relation between the expressions in the equation is presented from the qualitative aspect rather than the quantitative!
It shows that what is called the mass of a sub atomic particle (in this case the electron mass) is expressed by the relation between the 'orbital angular momentum' and the electric charge contained within this particle.
Another way to look at the electron mass m_e is thru the **magnetic flux quantum** Φ_0 expressed as:

$$e^2 = \left[\left(\frac{h}{2\pi}\right)^2\left(\frac{4\pi}{6m_e}\right)\right] \Rightarrow m_e = \underbrace{\left(\frac{h}{2e}\right)}_{\Phi_0}^2 \times 2/3\pi \Rightarrow m_e = \frac{2\times(\Phi_0)^2}{3\pi}$$

Note: The expression $\frac{h}{2e}$ is the inverse of the **Josephson constant** (denoted $\boldsymbol{K_J}$) and it represents the **magnetic flux quantum** Φ_0: $\Phi_0 \approx 2.0678\times10^{-15}[\mathrm{Wb}]$

8.1 The mass of a sub atomic particle is expressed by the relation between the 'orbital angular momentum' and the electric charge contained within this particle!

Or from another aspect:

8.2 The mass of a sub atomic particle is expressed as being a property related to the square of the magnetic flux quantum (magnetic flow rate that occurs within)!

$$m_{particle\ mass} = \mathrm{constant}\times\Phi_0^2\,[\mathrm{Wb}]^2$$

The other sub atomic particles' masses are a multiplication of the basic electrons' mass that results from the magnetic flux quantum property (magnetic flow rate) within them!

A new theoretical finding relating subatomic particles' mass to the square of the magnetic flux quantum provides a novel interpretation of mass and a novel interpretation of the Wave Function.

The formalism developed in this paper introduces a relationship between the masses of the electron, proton, and neutron and the square of the magnetic flux quantum. This relation was never known or probed. The Method used today to calculate the proton and neutron masses theoretically is based on the quantum chromodynamics theory of binding energy, which combines the kinetic energy of the quarks and the energy of the gluons within these particles, which requires a complicated calculation. The theory presented here calculates the electron, proton, and neutron masses in straightforward, nearly identical formulas, whose main component is the square of the magnetic flux quantum; the only difference between them is their Compton wavelength component, which is responsible for their different masses. Another formalism yields the proton's (hydrogen nucleus) and the neutron's radii directly from theory. The proton charge radius is a part of the actual proton radius. This theory presents a different approach to finding the proton's and the neutron's radii. It involves using universal constants such as Planck's mass and the universal gravitational constant, which ultimately yields a novel physical constant unfamiliar to science, in which the strong coupling constant in QCD is its derivative. Using this constant, it is possible to obtain the proton's radius. Combining the novel constant and the proton radius makes it possible to accurately calculate the Compton wavelength constants of the proton and the neutron, which are used to calculate their masses later. The theory also presents a novel way to describe the Planck mass and length and the gravitation constant through them. The gravitational constant is identified with Newton's law of universal gravitation. A new formula for the gravitational constant developed in this paper contains elements from the atomic domain (proton's mass and radius), which represent the quantum reality environment; in this way, they demonstrate the integration of the quantum and gravity levels. In Addition, the theory also presents a novel way for the Wave Function interpretation.

1. Relationships between the electron mass and the square of the magnetic flux quantum and between the Bohr radius and the vacuum permittivity.

The magnetic flux quantum Φ_0 is defined according to

(1) $$e = h/2\Phi_0 \rightarrow e^2 = h^2/4\Phi_0^2$$

where e is the elementary charge of an electron and h is Planck's constant. The electrostatic force acting on the electron at the Bohr level is

$$e^2/4\pi a_0^2 \varepsilon_0 = m_e v_e^2/a_0$$

(2) $$e^2 = m_e v_e^2 4\pi a_0 \varepsilon_0$$

where a_0 is the Bohr radius, ε_0 is the vacuum permittivity, m_e is the electron mass, and v_e is the electron velocity at the Bohr radius. The electron's angular momentum at the Bohr radius is

(3) $$\frac{h}{2\pi} = m_e v_e a_0 \rightarrow h = 2\pi m_e v_e a_0$$

By substituting Eq. (3) in Eq. (1) (the squared term), we obtain

$$e^2 = h^2/4\Phi_0^2$$

(4) $$e^2 = m_e v_e 2\pi a_0 \times m_e v_e 2\pi a_0/4\Phi_0^2$$

We can then rewrite Eq. (4) as follows:

(5) $$e^2 = m_e v_e^2 4\pi a_0 \times (\pi a_0 m_e/4\Phi_0^2)$$

By multiplying both the numerator and denominator of Eq. (5) by ε_0, we have

(6) $$e^2 = m_e v_e^2 4\pi a_0 \varepsilon_0 (\pi a_0 m_e/4\varepsilon_0 \Phi_0^2)$$

According to Eq. (2), the expression in parentheses in Eq. (6) should equal unity; thus, we obtain

(7) $$(\pi a_0 m_e/4\varepsilon_0 \Phi_0^2) = 1$$

We can then substitute the following values published by the National Institute of Standards and Technology Committee on Data for Science and Technology in 2018 (NIST CODATA 2018) in Eq. (7) (SI units):

$a_0 = 5.291772109 \times 10^{-11}$ m

$\varepsilon_0 = 8.8541878128 \times 10^{-12}\ \mathrm{A^2\,s^4\,kg^{-1}m^{-3}}$

$\Phi_0 = 2.067833848 \times 10^{-15}\ \mathrm{kg\,m^2\,s^{-2}\,A^{-1}}$

$\Phi_0^2 = 4.275936823 \times 10^{-30}\ \mathrm{kg^2\,m^4\,s^{-4}\,A^{-2}}$

$m_e = 9.1093837015 \times 10^{-31}\mathrm{kg}$

These substitutions give us the following relationship:

(8) $$\frac{\pi}{4} \times \frac{5.291772109 \times 10^{-11}\ \mathrm{m}}{8.8541878128 \times 10^{-12}\ \mathrm{A^2 s^4 kg^{-1} m^{-3}}} \times \frac{9.1093837015 \times 10^{-31}\ \mathrm{kg}}{4.275936823 \times 10^{-30}\ \mathrm{kg^2 m^4 s^{-4} A^{-2}}} = 1$$

Finally, we can consider the multiplication of the vacuum permittivity and the square of the magnetic flux in the denominator of Eq. (8). We can multiply their units:

(9) $$(\mathrm{A^2 s^4 kg^{-1} m^{-3}})(\mathrm{kg^2 m^4 s^{-4} A^{-2}}) = \mathrm{kg\,m}$$

We consider this reduction further in the next section.

2. Analysis of Equation (8).

a. After reducing the units in the denominator of Eq. (8), we obtained units corresponding to those in the numerator in Eq. (9). The result in Eq. (9) implies two options: Either Φ_0^2 has units of mass $[\mathrm{kg}]$ or units of length $[\mathrm{m}]$ or, vice versa, ε_0 has units of mass $[\mathrm{kg}]$ or has units of length $[\mathrm{m}]$.

b. The orders of magnitude in Eq. (8) indicate the scale of magnitude of these parameters. The Bohr radius scale is $\mathbf{10^{-11}}$m, and the vacuum permittivity scale is $\mathbf{10^{-12}}\ \mathrm{F\,m^{-1}}$. The scale of the electron mass is $\mathbf{10^{-31}}$kg, and the scale of the square of the magnetic flux quantum is $\mathbf{10^{-30}}[\mathrm{Wb}]^2$. From this consideration and the two options presented in (a), it is clearly nonsensical for ε_0 to have units of mass or Φ_0^2 to have units of length. No particles with a mass on the scale of $\mathbf{10^{-12}}$kg or a length on the scale of $\mathbf{10^{-30}}$m exist in the atom domain.

c. From these considerations, we can conclude with a high degree of certainty that the Bohr radius a_0 and the vacuum permittivity ε_0 are the same entity:

(10) $$\varepsilon_0 \equiv a_0$$

Moreover, the electron mass m_e and the square of the magnetic flux quantum Φ_0^2 are the same entity:

(11) $$m_e \equiv \Phi_0^2$$

d. The last conclusion is not as strange as it may seem, as the square of the magnetic flux quantum appears in the context of magnetic energy in a current loop and according to Albert Einstein's special theory of relativity, energy is equivalent to mass.

3. Using the conclusions from Sections 2(c) and 2(d) for the electron mass and vacuum permittivity.

By rearranging Eq. (7) for the electron mass calculation and substituting the values of $\varepsilon_0, a_0, \Phi_0^2$ from NIST CODATA 2018, we obtain

(12)

$$m_e = \frac{4}{\pi} \times \frac{\varepsilon_0}{a_0} \times \Phi_0^2 = \frac{4}{\pi} \times \frac{8.8541878128 \times 10^{-12}\mathrm{m}}{5.291772109 \times 10^{-11}\mathrm{m}} \times 4.275936823 \times 10^{-30}\mathrm{kg} = 9.109383697 \times 10^{-31}\mathrm{kg}$$

This value compares well with the NIST value of

$$m_e = 9.1093837015 \times 10^{-31}\mathrm{kg}$$

Another expression for m_e from Eq. (3) with the electron Compton wavelength $\lambda_e\,(v_e = c\alpha)$:

(13) $$\lambda_e = h / m_e c = \alpha 2\pi a_0$$

where c is the speed of light in vacuum and α is the fine structure constant. Here, we can rearrange Eq. (12) as $a_0 = 4\varepsilon_0 \Phi_0^2 \pi^{-1} m_e^{-1}$ and substitute this term in Eq. (13) for another expression of m_e:

(14) $$m_e = \left(\frac{8\,\alpha\,\varepsilon_0}{\lambda_e}\right)\Phi_0^2$$

To calculate the value of the vacuum permittivity ε_0, we rewrite Eq. (7) and substitute the values of Φ_0^2, m_e, a_0 from NIST CODATA 2018, which yields

(15) $$\varepsilon_0 = \frac{\pi}{4} \times \frac{m_e}{\Phi_0^2} \times a_0 = \frac{\pi}{4} \times \frac{9.1093837015 \times 10^{-31}\text{kg}}{4.275936823 \times 10^{-30}\text{kg}} \times 5.291772109 \times 10^{-11}\text{m} = 8.854187816 \times 10^{-12}\text{m}$$

This numerical value compares well with the NIST value of

$$\varepsilon_0 = 8.8541878128 \times 10^{-12}\ \text{A}^2\,\text{s}^4\,\text{kg}^{-1}\text{m}^{-3}.$$

4. Using the conclusions from Sections 2(c) and 2(d) for the elementary charge e and the Planck constant h.

In this subsection, we derive a new expression for the elementary charge e with $m_e = 4\varepsilon_0 \Phi_0^2 \pi^{-1} a_0^{-1}$ from Eq. (12) and with the fine structure constant α and the speed of light in vacuum c.

Here, we substitute m_e in Eq. (2) with $v_e^2 = c^2\alpha^2$ to obtain an expression of the elementary charge e:

$$e^2 = m_e c^2 \alpha^2 4\pi\varepsilon_0 a_0 = (4\varepsilon_0 \Phi_0^2 \pi^{-1} a_0^{-1})(c^2\alpha^2 4\pi\varepsilon_0 a_0)$$

(16) $$e = 4c\alpha\varepsilon_0\Phi_0$$

In Eq. (16), we substitute the values of ε_0, Φ_0, the speed of light in vacuum c, and the fine structure constant α from NIST CODATA 2018:

$$c = 2.99792458 \times 10^8\ \text{m}\,\text{s}^{-1}$$

$$\alpha = 7.2973525684 \times 10^{-3}$$

This substitution yields

(17) $$e = 1.602176633 \times 10^{-19}\ \text{C}$$

This value compares well with the NIST value of

$$e = 1.602176634 \times 10^{-19}\ \text{C}.$$

We can substitute $c^2\varepsilon_0 = \mu_0^{-1}$ (where μ_0 is the vacuum permeability) in Eq. (16) to obtain another expression for the elementary charge e:

(18) $$e = 4\alpha\Phi_0(\varepsilon_0/\mu_0)^{1/2}$$

By substituting the expression of e from Eq. (16) in Eq. (1), we obtain a new expression for h:

(19) $$h = 8c\alpha\varepsilon_0\Phi_0^2$$

By applying the values of $\varepsilon_0, \Phi_0, c, \alpha$ from NIST CODATA 2018 in Eq. (19), we find

$$h = 6.626070146 \times 10^{-34}\ \text{kg}\,\text{m}^2\text{s}^{-1}$$

This value compares well with the NIST value of $h = 6.626070146 \times 10^{-34}\ \text{kg}\,\text{m}^2\text{s}^{-1}$. Moreover, we obtain another expression for the Planck constant h by utilizing $c^2\varepsilon_0 = \mu_0^{-1}$:

(20) $$h = 8\alpha(\varepsilon_0/\mu_0)^{1/2}\Phi_0^2$$

5. Radius of the proton (hydrogen nucleus).

For further development, it is necessary to find the proton's radius. The nucleus of a hydrogen atom (proton) revolves around the center of mass shared with the electron. The rotation of both the electron and nucleus arises from considerations of momentum conservation in an isolated system and is taken into account by a computational correction called the reduced mass of the electron. The center of mass is very close to the axis of the nucleus because of its larger mass; thus, we can assume that the trajectory depicted by the nucleus while revolving around the center of mass lies at a distance almost equivalent to the nucleus radius. We will denote this radius as the proton radius, validated in the final result. As a side note, this radius is not equivalent to the proton's charge radius; however, there is a connection between these two parameters, which will be clarified in Section 5b. To find the proton's radius, we will use known formulas generated for the natural units of the Stoney and Planck scales. We will start with the Stoney

scale, from which we will move to the Planck scale.
The Stoney length l_s in natural units is the following

(21) $$l_s = \sqrt{\frac{e^2 G}{4\pi\varepsilon_0 c^4}}$$

where G is the gravitational constant. The Stoney mass m_s from the natural units is

(22) $$m_s = \sqrt{\frac{e^2}{4\pi\varepsilon_0 G}}$$

Or rewrite Eq. (22) for the gravitational constant G:

(23) $$G = \frac{e^2}{4\pi\varepsilon_0 m_s^2}$$

Substituting the relation $e^2 = 2hc\alpha\varepsilon_0$ from the definition of the fine structure constant in Eq. (23),

(24) $$G = \frac{2hc\alpha\varepsilon_0}{4\pi\varepsilon_0 m_s^2}$$

The orbital angular momentum of the proton at the trajectory around the center of mass should be expressed by the reduced Planck constant.
The proton's velocity at this trajectory is denoted here as v_p. An initial estimation of the velocity v_p yields approximately one fifth of the speed of light in vacuum. Hence, it is necessary to add a relativistic element $(1-v_p^2/c^2)^{1/2}$ with $v_p = c\beta$:

$$\frac{h}{2\pi} = m_p v_p r_p \sqrt{1-v_p^2/c^2}$$

(25) $$h = m_p c\beta 2\pi r_p \sqrt{1-\beta^2}$$

where m_p is the proton mass, β is the ratio of v_p to *c*, and r_p is the proton radius. By substituting the expression of h from Eq. (25) in Eq. (24) and reducing the expression, we obtain

(26) $$G = \frac{m_p c^2 \alpha\beta r_p \sqrt{1-\beta^2}}{m_s^2}$$

The β constant is similar to the fine structure constant α, also known as the electromagnetic coupling constant, that also appears in the electron's velocity expression at the Bohr radius as $v_e = c\alpha$.
We can divide Eq. (21) by Eq. (22) ($4\pi\varepsilon_0$ is reduced, and the elementary charge e is partially reduced):

(27) $$\frac{l_s}{m_s} = \sqrt{\frac{e^2}{e^2} \times \frac{G^2}{c^4}} = \frac{e}{e} \times \frac{G}{c^2}$$

Then rearrange Eq. (27) to obtain an expression for G:

(28) $$G = \frac{e}{e} \times \frac{l_s}{m_s} c^2$$

By setting the expressions in Eq. (28) and Eq. (26) equal to each other, we have

(29) $$\frac{e}{e} \times \frac{l_s}{m_s} c^2 = \frac{m_p \alpha\beta c^2 r_p \sqrt{1-\beta^2}}{m_s^2}$$

We then divide both sides of Eq. (29) by (e/e), multiply both sides by m_s^2, reduce, and rearrange:

(30) $$m_s \times l_s \times c^2 = \left[\left(\frac{1}{e} m_p \beta\right)\left(\frac{\alpha\, r_p \sqrt{1-\beta^2}}{\frac{1}{e}}\right)\right] \times c^2$$

Eq. (30) presents a similarity between the right and left flanks (mass component and length component). The expression is split into two parts on the right-hand side of the equation because it contains the solutions corresponding to actual experimental results in the final analysis.

The following new expressions are proposed solutions for the Stoney units.

New expression of Stoney mass:

$$m_s = \frac{1}{e} m_p \beta$$

New expression of Stoney length:

$$l_s = \left(\frac{\alpha\, r_p \sqrt{1-\beta^2}}{\frac{1}{e}} \right)$$

Note that the $(1/e)$ expression in Eq. (30) represents a dimensionless number, for instance, the number of charged particles in one Coulomb $[\mathrm{C}]$ unit.

(31) $$\frac{1}{e} = \frac{1\not{C}}{1.602176634\times 10^{-19}\not{C}} = 6.2415090744\times 10^{18}$$

This number, as a multiplier, creates a quantity of charged particles (in our case, the number of protons contained within the Planck mass, which corresponds to a quintillion protons) or, as a divisor, creates the smallest length (in our case, a contracted radius of the proton within the Planck mass under internal attraction forces, which corresponds to a quintillionth of the proton radius that represents the Planck length). This expression is displayed in the following equations, as in Eq. (31).

The gravitational constant G in Stoney units from Eq. (28) with the proposed new expressions is:

(32) $$G = \frac{\left(\frac{\alpha\, r_p \sqrt{1-\beta^2}}{\frac{1}{e}} \right)}{\left(\frac{1}{e} m_p \beta \right)} \times c^2$$

We can then set Eq. (23) and Eq. (32) equal to each other and substitute the square of the Stoney mass term $m_s^2 = \left(\frac{1}{e} m_p \beta \right)^2$ in the denominator of Eq. (23) as:

(33) $$\frac{e^2}{4\pi\varepsilon_0 \left(\frac{1}{e} m_p \beta \right)^2} = \frac{\alpha\, r_p \sqrt{1-\beta^2}}{\frac{1}{e}\left(\frac{1}{e} m_p \beta \right)} \times c^2$$

Multiplying both sides of Eq. (33) by $4\pi\varepsilon_0 m_p \beta$, reducing, and rearranging yields

(34) $$e^2 = 2\left[m_p c \beta 2\pi r_p \sqrt{1-\beta^2} \right] c\alpha\varepsilon_0$$

The expression of Eq. (34) shows the equivalence of $e^2 = 2hc\alpha\varepsilon_0$, where the right-hand side (in brackets) contains the expression of the Planck constant h with the proton parameters introduced in Eq. (25). This result confirms the choice of the proposed solutions for the Stoney units of mass and length from Eq. (30). Although this option was based on a logical consideration, there are additional combinations that could be chosen that yield incorrect results.

We multiply the numerator and denominator of Eq. (32) by $\alpha^{-1/2}$ to obtain the G at the Planck scale:

(35) $$G = \frac{\left(\frac{\alpha^{1/2}\, r_p \sqrt{1-\beta^2}}{\frac{1}{e}} \right)}{\frac{1}{\alpha^{1/2}}\left(\frac{1}{e} m_p \beta \right)} \times c^2$$

The difference between the Stoney and Planck units arises from the need to multiply Planck units by the square root of the fine structure constant, $\alpha^{1/2}$. Consequently, we obtain the following expressions.
New expression of Planck mass:

$$m_{\text{pl}} = \frac{1}{\alpha^{1/2}}\left(\frac{1}{e}m_p\beta\right)$$

New expression of Planck length:

$$l_{\text{pl}} = \left(\frac{\alpha^{1/2}\, r_p\sqrt{1-\beta^2}}{\frac{1}{e}}\right)$$

By using the Planck mass in natural units and the new expression of Planck mass, we can derive the expression and value of β. The Planck mass, defined by natural units, is

(36) $$m_{\text{pl}} = \sqrt{\frac{hc}{2\pi G}}$$

The new expression for the Planck mass from Eq. (35) is

(37) $$m_{\text{pl}} = \frac{1}{\alpha^{1/2}}\left(\frac{1}{e}m_p\beta\right)$$

We set Eq. (36) and Eq. (37) as equal:

(38) $$\sqrt{\frac{hc}{2\pi G}} = \frac{1}{\alpha^{1/2}}\left(\frac{1}{e}m_p\beta\right)$$

Then rearrange Eq. (38) to obtain an expression for β:

(39) $$\beta = \sqrt{\frac{hc\alpha}{2\pi G}} \times \frac{1}{1/e \times m_p}$$

In Eq. (39), we substitute the values of h, c, α, m_p and the following values from NIST CODATA 2018:

$1/e = 6.2415090744 \times 10^{18}$

$G = 6.6743 \times 10^{-11}\text{m}^3\,\text{kg}^{-1}\text{s}^{-2}$

We obtain $\beta = 0.178090537$

As a side note, the relationship of β to nuclear research is through the strong coupling constant in QCD, which is a β derivative: $\alpha_s = 2/3\,\beta \rightarrow \alpha_s = 2/3 \times 0.178090537 = 0.118727$

This value compares well with the value obtained experimentally, $\alpha_s \approx 0.1187$

Using the Planck length from natural units and the new expression for the Planck length, we can derive the expression and value of the proton radius r_p.

(40) $$l_{\text{pl}} = \sqrt{\frac{hG}{2\pi c^3}}$$

The new expression for the Planck length from Eq. (35) is

(41) $$l_{\text{pl}} = \left(\frac{\alpha^{1/2}\, r_p\sqrt{1-\beta^2}}{\frac{1}{e}}\right)$$

By setting the expressions in Eq. (40) and Eq. (41) expressions equal, we obtain

(42) $$\sqrt{\frac{hG}{2\pi c^3}} = \left(\frac{\alpha^{1/2}\, r_p\sqrt{1-\beta^2}}{\frac{1}{e}}\right)$$

We then rearrange Eq. (42) for the proton radius r_p:

(43) $$r_p = \sqrt{\frac{hG}{2\pi c^3 \alpha}} \times \frac{1}{e} \times \frac{1}{\sqrt{1-\beta^2}}$$

substituting $\beta = 0.178090537$ and the values of h, G, c, α and $1/e$ from NIST CODATA 2018 in Eq. (43) yields $r_p = 1.200094665 \times 10^{-15}\,\text{m}$

This value compares well with the value obtained experimentally, explained below:

a. The proton radius obtained in Eq. (43) complies with the experimental formulation that assumes a spherical nucleus with radius expressed by the Fermi equation for the nuclear radius R_n: $R_\text{n} = R_0 \text{A}^{1/3}$, where R_0 is an essamtion made from experimental results to be $R_0 = 1.2 \times 10^{-15}\,\text{m}$, and A is the atomic number. For hydrogen, $\text{A}=1$ and $R = 1.2 \times 10^{-15}\,\text{m}$.

The example for the proton charge radius in the following section is presented without overall proof, which requires a separate article.

b. The proton charge radius (r_{pcr}) represents the maximum distance from the proton axis that the electron or muon reaches in their penetration to the proton due to interactions with up quarks. This radius is expressed as

$$r_{pcr} = 4\beta \times r_p (1-\beta^2)^{1/2}$$

Substituting the following in the expression for r_{pcr} $r_p = 1.200094665 \times 10^{-15}\,\text{m}$ and $\beta = 0.178090537$, we obtain $r_{pcr} = 8.4123564 \times 10^{-16}\,\text{m}$

This value is similar to the NIST value of $r_{pcr} = 8.414 \times 10^{-16}\,\text{m}$.

The proton's Compton wavelength from Eq. (25) is

(44) $$\lambda_p = \frac{h}{m_p c} = 2\pi\beta r_p \sqrt{1-\beta^2}$$

Substituting the values of β and r_p in Eq. (44) yields

$$\lambda_p = 2\pi \times 0.178090537 \times 1.200094665 \times 10^{-15}\,\text{m} \times 0.984014106 = 1.3214098539 \times 10^{-15}\,\text{m}$$

This values compares well with the NIST value of

$$\lambda_p = 1.32140985539 \times 10^{-15}\,\text{m}.$$

The last result shows that β combined with the proton radius r_p obtained from Eq. (43) and used in Eq. (44) is entirely consistent with the value of the proton's
Compton wavelength λ_p from NIST CODATA 2018, confirming the validity of our approach.

To obtain the gravitational constant G, we utilize Eq. (32) and substitute the m_p, c, α from NIST CODATA 2018 and also $1/e = 6.2415090744 \times 10^{18}$, β, and r_p:

(45) $$G = \frac{\left(\dfrac{r_p\sqrt{1-\beta^2}}{6.2415090744 \times 10^{18}}\right)\alpha}{\left(6.2415090744 \times 10^{18} \times m_p\right)\beta} \times c^2 = 6.6743 \times 10^{-11}\,\text{m}^3\text{kg}^{-1}\text{s}^{-2}$$

The last result compares well with the NIST value of

$$G = 6.6743 \times 10^{-11}\,\text{m}^3\text{kg}^{-1}\text{s}^{-2}.$$

6. Radius of the neutron.

The ratio between the proton mass and neutron mass is the same as the ratio between the neutron Compton wavelength and proton Compton wavelength. The values of $m_p, m_n, \lambda_n, \lambda_p$ are substituted from the NIST CODATA 2018 in the following ratios:

$$\frac{m_p}{m_n} = \frac{1.67262192369\times10^{-27}\,\text{Kg}}{1.67492749804\times10^{-27}\,\text{Kg}} = 0.998623478$$

$$\frac{\lambda_n}{\lambda_p} = \frac{1.31959090581\times10^{-15}\,\text{m}}{1.32140985539\times10^{-15}\,\text{m}} = 0.998623478$$

This result indicates that the ratio is also appropriate for the ratio between the neutron and proton radii.

(46) $$\frac{m_p}{m_n} = \frac{r_n}{r_p} \rightarrow r_n = \frac{m_p}{m_n}\times r_p$$

By substituting the values of m_p, m_n from NIST CODATA 2018 and the radius r_p in Eq. (46) for the neutron radius, we obtain

$$r_n = \frac{1.67262192369\times10^{-27}\,\text{kg}}{1.67492749804\times10^{-27}\,\text{kg}}\times1.200094667\times10^{-15}\,\text{m} = 1.19844271\times10^{-15}\,\text{m}$$

The proton and neutron are almost identical in size, and the β constant is related to both radii. Consequently, $v_n = v_p = c\beta$ .This result is validated by the neutron Compton wavelength λ_n with β and r_n

(47) $$\lambda_n = \frac{h}{m_n c} = 2\pi\beta r_n\sqrt{1-\beta^2}$$

Substituting the values of β and r_n in Eq. (47) for λ_n, we obtain

$$\lambda_n = 2\pi\times0.178090537\times1.19844271\times10^{-15}\,\text{m}\times0.984014105 = 1.31959090471\times10^{-15}\,\text{m}$$

This value matches well with the NIST value of

$$\lambda_n = 1.31959090581\times10^{-15}\,\text{m}.$$

The result obtained by combining β with the neutron radius r_n given by Eq. (46) and substituting in Eq. (47) is entirely consistent with the NIST CODATA 2018 value of the neutron Compton wavelength λ_n, confirming the validity of $v_n = v_p = c\beta$.

7. Additional expressions for the proton and neutron masses and radii.

We divide Eq. (3) (with $v_e = c\alpha$) by Eq. (25) as follows:

(48) $$\frac{m_e 2\pi\alpha a_0}{m_p 2\pi\beta\, r_p\sqrt{1-\beta^2}} = 1$$

By rearranging Eq. (48) and solving for the proton mass, we obtain

(49) $$m_p = \frac{m_e \alpha a_0}{\beta\times r_p\sqrt{1-\beta^2}}$$

Substituting $m_e = 4\varepsilon_0\Phi_0^2\,\pi^{-1}a_0^{-1}$ from Eq. (12) in Eq. (49) yields the following expression for m_p :

(50) $$m_p = \frac{4}{\pi}\times\frac{\alpha}{\beta}\times\frac{\varepsilon_0}{r_p\sqrt{1-\beta^2}}\times\Phi_0^2$$

Substituting $\lambda_e = 2\pi\alpha a_0$ and $\lambda_p = 2\pi\beta r_p\sqrt{1-\beta^2}$ in Eq.(48) yields

(51) $$\frac{m_e}{m_p} = \frac{\lambda_p}{\lambda_e} \quad \text{Or} \rightarrow m_p = \frac{m_e\lambda_e}{\lambda_p}$$

We substitute $\lambda_e = \alpha 2\pi a_0$ and m_e from Eq. (12) in Eq. (49) to obtain another expression for m_p :

(52) $$m_p = \left(\frac{8\,\alpha\,\varepsilon_0}{\lambda_p}\right)\Phi_0^2$$

Applying the NIST CODATA 2018 value for the proton mass m_p in Eq. (52) gives

(53)

$$m_p = \frac{8\times 7.297352569\times 10^{-3}\times 8.8541878128\times 10^{-12}\,\text{m}}{1.321409\,85539\times 10^{-15}\,\text{m}}\times 4.27593682\times 10^{-30}\,\text{kg} = 1.6726219217\times 10^{-27}\,\text{kg}$$

This value matches well with the NIST value of

$$m_p = 1.67262192369\times 10^{-27}\,\text{kg}.$$

We can rearrange Eq. (47) to obtain the orbital angular momentum of the neutron:

(54) $$\frac{h}{2\pi} = m_n c\beta r_n\sqrt{1-\beta^2} \text{ Or } h = m_n c\beta 2\pi r_n\sqrt{1-\beta^2}$$

We then divide Eq. (3) by Eq. (54), as follows:

(55) $$\frac{m_e c\alpha 2\pi a_0}{m_n c\beta\, 2\pi r_n\sqrt{1-\beta^2}} = 1$$

Rearranging Eq. (55) for the neutron mass, gives

(56) $$m_n = \frac{m_e\alpha a_0}{\beta\, r_n\sqrt{1-\beta^2}}$$

Substituting m_e from Eq. (12) in Eq. (56) yields an expression for the neutron mass m_n:

(57) $$m_n = \frac{4}{\pi}\times\frac{\alpha}{\beta}\times\frac{\varepsilon_0}{r_n\sqrt{1-\beta^2}}\times\Phi_0^2$$

Substituting $\lambda_e = 2\pi\alpha a_0$ from Eq. (13) and $\lambda_n = 2\pi\beta r_n\sqrt{1-\beta^2}$ from Eq. (47) in Eq. (55) yields

(58) $$\frac{m_e}{m_n} = \frac{\lambda_n}{\lambda_e} \quad \text{Or} \rightarrow m_n = \frac{m_e\lambda_e}{\lambda_n}$$

We substitute $\lambda_e = \alpha 2\pi a_0$ and $m_e = 4\varepsilon_0\Phi_0^2\pi^{-1}a_0^{-1}$ from Eq. (12) in Eq. (58) for another expression of the neutron mass m_n:

(59) $$m_n = \left(\frac{8\,\alpha\,\varepsilon_0}{\lambda_n}\right)\Phi_0^2$$

Substituting the NIST CODATA 2018 value for the neutron mass m_n in Eq. (59) yields

(60)

$$m_n = \frac{8\times 7.2973525693\times 10^{-3}\times 8.8541878128\times 10^{-12}\,\text{m}}{1.319\,59090581\times 10^{-15}\,\text{m}}\times 4.275936823\times 10^{-30}\,\text{kg} = 1.674927497\times 10^{-27}\,\text{kg}$$

This value compares well with the NIST value of

$$m_n = 1.67492749804\times 10^{-27}\,\text{kg}.$$

The ratio of proton mass to electron mass is obtained from Eq. (49) after rearranging and substituting the NIST CODATA 2018 values for α, a_0 and β and r_p;

(61) $$m_p/m_e = \alpha/\beta\times a_0/r_p\sqrt{1-\beta^2} = 1836.152675$$

This ratio can also be obtained from Eq. (51). This value matches well with the NIST ratio of

$$m_p/m_e = 1836.1526734.$$

The ratio of neutron mass to electron mass is obtained from Eq. (56) after rearranging and substituting NIST CODATA 2018 values for α, a_0 and β and r_p;

(62) $$m_n/m_e = \alpha/\beta\times a_0/r_n\sqrt{1-\beta^2} = 1838.683661$$

This value matches well with the NIST ratio of

$$m_n/m_e = 1838.683661$$

8. The squared values of the magnetic flux quantum used in the wave function, yield solutions which depict the flow pattern of the magnetic flux surrounding electrons at a given energy level

Using the Normalized Wave Function of Hydrogen Atom equation of L. Schiff from his book The Quantum Nechanics, 2nd ed' (New York: McGraw-Hill,1968),page 93, as follows

(63) $$R_{nl}(r)=\sqrt{\left(\frac{2Z}{na_0}\right)^3\frac{(n-1-l)!}{2n[(n+l)!]^3}}\left(\frac{2Zr}{na_0}\right)^l e^{-zr/na_0}L_{n+l}^{2l+1}\left(\frac{2Zr}{na_0}\right)$$

Substituting $a_0 = 4\varepsilon_0\Phi_0^2\pi^{-1}m_e^{-1}$ a rewrite of Eq. (7) in Eq. (63), and after rearranging, it gives

(64) $$R_{nl}(r)=\left(\frac{\pi r}{2\varepsilon_0}\times\frac{zm_e}{n\Phi_0^2}\right)^{3/2}\left(\frac{(n-1-l)!}{2n[(n+l)!]^3}\right)^{1/2}\left(\frac{\pi r}{2\varepsilon_0}\times\frac{zm_e}{n\Phi_0^2}\right)^l e^{-zr/na_0}L_{n+l}^{2l+1}\left(\frac{\pi r}{2\varepsilon_0}\times\frac{zm_e}{n\Phi_0^2}\right)$$

Now using the third term in Eq. (64) as following

(65) $$\left(\frac{\pi r}{2\varepsilon_0}\times\frac{zm_e}{n\Phi_0^2}\right)^l$$

As an example it provides the value of the derivative r at the Electron Wave Function of Hydrogen like Atom equation, as following

(66) $$r=\frac{2\varepsilon_0}{\pi}\times\frac{n\Phi_0^2}{zm_e}\quad \text{When } l=0\rightarrow\left(\frac{\pi r}{2\varepsilon_0}\times\frac{zm_e}{n\Phi_0^2}\right)^0$$

Conclusions.

a. The conclusions from Eq. 64 for the electron wave function of 'Hydrogen like Atom' $R_{nl}(r)$ that it is mainly a function of the changing value of the square of the magnetic flux quantum $n\Phi_0^2$ and the number of electrons zm_e at the relevant level (Please notice that the number of electrons in a neutral atom is equal to the number of protons in the nucleus of the atom presented as the atomic number z).

b. The electron wave function in a 'Hydrogen-like atom' depends on the radius r of the atomic level expressed by several Bohr radii, and since the Bohr radius is a function of the square of the magnetic flux quantum Φ_0^2, the electron wave function describes the magnitude of the flow pattern of the magnetic flux that surrounds the electrons at the energy levels in the atom.

c. The mass of the electron and other subatomic particles is related to the magnitude of the square of the magnetic flux quantum, which makes up the particles. This relationship results in a novel expression of universal constants.

d. The Gravitational constant is identified based on Newton's law of universal gravitation. The new formula for the Gravitational constant developed in this book contains elements from the atomic domain (proton's mass and radius), which represent the quantum reality environment; in this way, they demonstrate the integration of the quantum and gravity levels.

The Quantum numbers of Schrodinger equation obtained thru different approach.

* In order to speak one language, the symbols used in solving the quantum numbers of Schrödinger equation are used in this new approach that provides a broader understanding of their meaning!

1. The Principal number **n**.

The Principal number n defines an energy level that has a capacity of a maximum number of electrons that their allocation is dictated by their energy and their angular momentum criterion. The number of electrons at a specific energy level can be determined by examining their angular momentum jointly with the Principal number in the following process:

Substitute Eq. 242 here:

$$\vec{L}_{(n)} = n(m_e \times \vec{v}_{e(n)} \times r_n)$$

In Eq. 241 here:

$$e^2 = \vec{L}_n^2\left(\frac{4\pi}{6m_e}\right)$$

Yields the following:

$$e^2 = n(m_e \times \vec{v}_{e(n)} \times r_n) \times n(m_e \times \vec{v}_{e(n)} \times r_n) \times (4\pi / 6m_e)$$

And Substitute the following in the last equation:

$$\vec{\hbar} = (m_e \times \vec{v}_{e(n)} \times r_n) \rightarrow \vec{L}_{(n)} = n\vec{\hbar}$$

It yields the following equation:

$$e^2 = n(\vec{\hbar}) \times n(\vec{\hbar}) \times (4\pi / 6m_e)$$

And by reducing m_e from one of $\vec{\hbar}$ it yields the required term following here:

$$e^2 = n^2 \times (\pm\vec{\hbar}) \times [(4\pi / 6) \times (r_n) \times (\pm\vec{v}_n)]$$

Note 1: The $\vec{\hbar}$ (OAM) is a vector depending as such on the velocity $\vec{v}_n$ which either can be positive or negative as per revolving direction!

We use the last expression in two complementary parts. One presented with $(-\vec{\hbar})$ and a velocity $(-\vec{v}_n)$ and the other with $(+\vec{\hbar})$ and a velocity $(+\vec{v}_n)$ and then combining both for complete required results.

Note 2: The velocity of an electron $(\vec{v}_n)$ at each energy level is a function of the level's radius r_n so the expression in the square brackets in the last equation remains constant. The electron's velocity $(-\vec{v}_n)$ as a vector property dictates the direction sign of the relevant orbital angular momentum $(-\vec{\hbar})$.

1a. $e^2 = n^2 \times (-\vec{\hbar}) \times [(4\pi / 6) \times (r_n) \times (-\vec{v}_n)]$ For negative option of $(\vec{\hbar})$

$n = 1 \rightarrow e^2 = \mathbf{1}^2 \times (-\vec{\hbar}) \times [(4\pi / 6) \times (r_1) \times (-\vec{v}_1)] \rightarrow e^2 = \mathbf{1}(-\vec{\hbar}) \times [(4\pi / 6) \times (r_1) \times (-\vec{v}_1)]$

$n = 2 \rightarrow e^2 = \mathbf{2}^2 \times (-\vec{\hbar}) \times [(4\pi / 6) \times (r_2) \times (-\vec{v}_2)] \rightarrow e^2 = \mathbf{4}(-\vec{\hbar}) \times [(4\pi / 6) \times (r_2) \times (-\vec{v}_2)]$

$n = 3 \rightarrow e^2 = \mathbf{3}^2 \times (-\vec{\hbar}) \times [(4\pi / 6) \times (r_3) \times (-\vec{v}_3)] \rightarrow e^2 = \mathbf{9}(-\vec{\hbar}) \times [(4\pi / 6) \times (r_3) \times (-\vec{v}_3)]$

$n = 4 \rightarrow e^2 = \mathbf{4}^2 \times (-\vec{\hbar}) \times [(4\pi / 6) \times (r_4) \times (-\vec{v}_4)] \rightarrow e^2 = \mathbf{16}(-\vec{\hbar}) \times [(4\pi / 6) \times (r_4) \times (-\vec{v}_4)]$

$n = 5 \rightarrow e^2 = \mathbf{5}^2 \times (-\vec{\hbar}) \times [(4\pi / 6) \times (r_5) \times (-\vec{v}_5)] \rightarrow e^2 = \mathbf{25}(-\vec{\hbar}) \times [(4\pi / 6) \times (r_5) \times (-\vec{v}_5)]$

1b. $e^2 = \mathbf{n}^2 \times (+\vec{\hbar}) \times [(4\pi / 6) \times (r_n) \times (+\vec{v}_n)]$ For positive option of $(\vec{\hbar})$

$n = 1 \rightarrow e^2 = \mathbf{1}^2 \times (+\vec{\hbar}) \times [(4\pi / 6) \times (r_1) \times (+\vec{v}_1)] \rightarrow e^2 = \mathbf{1}(+\vec{\hbar}) \times [(4\pi / 6) \times (r_1) \times (+\vec{v}_1)]$

$n = 2 \rightarrow e^2 = \mathbf{2}^2 \times (+\vec{\hbar}) \times [(4\pi / 6) \times (r_2) \times (+\vec{v}_2)] \rightarrow e^2 = \mathbf{4}(+\vec{\hbar}) \times [(4\pi / 6) \times (r_2) \times (+\vec{v}_2)]$

$n = 3 \rightarrow e^2 = \mathbf{3}^2 \times (+\vec{\hbar}) \times [(4\pi / 6) \times (r_3) \times (+\vec{v}_3)] \rightarrow e^2 = \mathbf{9}(+\vec{\hbar}) \times [(4\pi / 6) \times (r_3) \times (+\vec{v}_3)]$

$n = 4 \rightarrow e^2 = \mathbf{4}^2 \times (+\vec{\hbar}) \times [(4\pi / 6) \times (r_4) \times (+\vec{v}_4)] \rightarrow e^2 = \mathbf{16}(+\vec{\hbar}) \times [(4\pi / 6) \times (r_4) \times (+\vec{v}_4)]$

$n = 5 \rightarrow e^2 = \mathbf{5}^2 \times (+\vec{\hbar}) \times [(4\pi / 6) \times (r_5) \times (+\vec{v}_5)] \rightarrow e^2 = \mathbf{25}(+\vec{\hbar}) \times [(4\pi / 6) \times (r_5) \times (+\vec{v}_5)]$

Following here are the maximum numbers of electrons at a specific level thru their angular momentum (OAM) from subsections **1a. 1b**.

1c. The relationships $e^2 \propto n^2(-\vec{\hbar})$ and $e^2 \propto n^2(+\vec{\hbar})$ express the relation between the particle's electric charge and its (OAM).Their total sum determines the maximum electrons allocated at level n.

$n=1 \rightarrow\ e^2 \propto \mathbf{1}(-\vec{\hbar})$ And $e^2 \propto \mathbf{1}(+\vec{\hbar}) \rightarrow \underbrace{\mathbf{1}(-\vec{\hbar}) \text{ and } \mathbf{1}(+\vec{\hbar})}_{\text{Total of } \mathbf{2} \text{ electrons}}$

$n=2 \rightarrow\ e^2 \propto \mathbf{4}(-\vec{\hbar})$ And $e^2 \propto \mathbf{4}(+\vec{\hbar}) \rightarrow \underbrace{\mathbf{4}(-\vec{\hbar}) \text{ and } \mathbf{4}(+\vec{\hbar})}_{\text{Total of } \mathbf{8} \text{ electrons}}$

$n=3 \rightarrow\ e^2 \propto \mathbf{9}(-\vec{\hbar})$ And $e^2 \propto \mathbf{9}(+\vec{\hbar}) \rightarrow \underbrace{\mathbf{9}(-\vec{\hbar}) \text{ and } \mathbf{9}(+\vec{\hbar})}_{\text{Total of } \mathbf{18} \text{ electrons}}$

$n=4 \rightarrow\ e^2 \propto \mathbf{16}(-\vec{\hbar})$ And $e^2 \propto \mathbf{16}(+\vec{\hbar}) \rightarrow \underbrace{\mathbf{16}(-\vec{\hbar}) \text{ and } \mathbf{16}(+\vec{\hbar})}_{\text{Total of } \mathbf{32} \text{ electrons}}$

$n=5 \rightarrow\ e^2 \propto \mathbf{25}(-\vec{\hbar})$ And $e^2 \propto \mathbf{25}(+\vec{\hbar}) \rightarrow \underbrace{\mathbf{25}(-\vec{\hbar}) \text{ and } \mathbf{25}(+\vec{\hbar})}_{\text{Total of } \mathbf{50} \text{ electrons}}$

2. The orbital quantum number operator $\boldsymbol{L}$:

$\vec{L} = \sqrt{l(l+1)} \times \vec{\hbar}$ Where l is the azimuthal orbital number $l = (n-1)$ and $\vec{\hbar}$ is the reduced Planck constant express the Orbital Angular Momentum (OAM) of an electron.

Substitute $l = (n-1)$ in $\vec{L} = \sqrt{l(l+1)} \times \vec{\hbar}$ yields $L = \sqrt{l \times n} \times \vec{\hbar}$ or $L^2 = (l \times \vec{\hbar})(n \times \vec{\hbar}) \rightarrow$

The Azimuthal number l can be understood as a previous energy level when saying that $l = (n-1)$.

* The orbital quantum number operator $\vec{L} = \sqrt{l \times n} \times \vec{\hbar}$ can be interpreted as the <u>geometrical mean</u> of the product of the relevant energy level n and l the orbital that comes previous to it $\rightarrow l = (n-1)$!

** The <u>mean value</u> of $\vec{L}$ (the orbital quantum number operator) is linked to the higher energy level!

Examples:

$n = 1 \rightarrow \underbrace{(1-1=0)}_{\text{no orbitals prior to } n=1} \rightarrow l = 0, L^2 = \underbrace{(0 \times \vec{\hbar})}_{\text{at } n-1} \times \underbrace{(1 \times \vec{\hbar})}_{\text{at } n=1} = 0$; The <u>mean OAM</u> of an electron at n = 1 is zero, not its actual!

Note: The meaning of this is that the <u>mean OAM</u> designated as $\vec{L}$ <u>is equal to zero</u> because of the value of l (the Azimuthal number) in the product above which refers to a level that actually doesn't exist (a level which should come before $n=1$) and hence it's been expressed as equal to zero, but it lets the electron acquire an OAM at $n=1$ (the zero OAM obtained at the experiments is because of the interaction between the proton's total inner OAM ($+\vec{\hbar}$) and the electron's ($-\vec{\hbar}$) at ground state!).

$n = 2 \rightarrow \underbrace{l = (2-1)}_{\text{previous level is } n=1} \rightarrow l = 1 \rightarrow \vec{L} = \pm\sqrt{1 \times 2} \times \vec{\hbar}$ The <u>mean OAM</u> of an electron at n = 2 is $\vec{L} = \pm\sqrt{2} \times \vec{\hbar}$

$n = 3 \rightarrow \underbrace{l = (3-1)}_{\text{previous level is } n=2} \rightarrow l = 2 \rightarrow \vec{L} = \pm\sqrt{2 \times 3} \times \vec{\hbar}$ The <u>mean OAM</u> of an electron at n = 3 is $\vec{L} = \pm\sqrt{6} \times \vec{\hbar}$

$n = 4 \rightarrow \underbrace{l = (4-1)}_{\text{previous level is } n=3} \rightarrow l = 3 \rightarrow \vec{L} = \pm\sqrt{3 \times 4} \times \vec{\hbar}$ The <u>mean OAM</u> of an electron at n = 4 is $\vec{L} = \pm\sqrt{12} \times \vec{\hbar}$

$n = 5 \rightarrow \underbrace{l = (5-1)}_{\text{previous level is } n=4} \rightarrow l = 4 \rightarrow \vec{L} = \pm\sqrt{4 \times 5} \times \vec{\hbar}$ The <u>mean OAM</u> of an electron at n = 5 is $\vec{L} = \pm\sqrt{20} \times \vec{\hbar}$

Note: The mean value of the orbital angular momentum (OAM) between two energy levels occurs because the electrons at each two close levels interact with each other, and this is also the reason that splits the so called atom's shells into sub shells.

3. The Magnetic quantum number m_l

The orbital quantum number operator L which is value express the mean orbital angular momentum (OAM) between two close energy levels also splits the so called atom's shells into a symmetrical sub shells around a relevant energy level designated by the principal number n . The Magnetic quantum number m_l indicates the angular momentum (OAM) value of electrons in sub shells that has a non-zero (OAM) component on the ($\pm z$) axis. If L has no projection on the ($\pm z$) axis, it means that a relevant (OAM) is perpendicular to ($\pm z$) axis (though the actual OAM is not zero!)

The levels above and below a relevant energy level n which is affected by the so called orbital quantum number operator L, are for it a kind of analogy to orbital sub shells and as such they constitute a basis for calculating the actual sub shells of the relevant energy level. It is done by subtraction of the relevant principal number n that represents an energy level from its previous and than from its next and including itself at the time. Further will continue to next energy level, but each time will consist of one more levels to be subtracted from the relevant one which shows a symmetric form of negative and positive OAM values, that represents the electrons mutual attraction requirements while being allocated at the specific sub shell! Following here is how it is done:

Note: The zero value express an opposed OAM vectors on plane x-y that are perpendicular to z axis!

$$n=1 \rightarrow \quad \underbrace{(1-0)}_{\substack{\text{no orbitals}\\ \text{prior to } n=1}} \quad \underbrace{(1-1)}_{\text{same level}} \quad \underbrace{(1-2)}_{\substack{\text{next level}\\ \text{is } n=2}} \rightarrow \underbrace{\mathbf{1,0,-1}}_{m_l \text{ numbers}}$$

$$n=2 \rightarrow \quad \underbrace{(2-1)}_{\substack{\text{previous level}\\ \text{is } n=1}} \quad \underbrace{(2-2)}_{\text{same level}} \quad \underbrace{(2-3)}_{\substack{\text{next level}\\ \text{is } n=3}} \rightarrow \underbrace{\mathbf{1,0,-1}}_{m_l \text{ numbers}}$$

$$n=3 \rightarrow \quad \underbrace{(3-1),(3-2)}_{\substack{\text{previous levels}\\ \text{are } n=1,\, n=2}} \quad \underbrace{(3-3)}_{\text{same level}} \quad \underbrace{(3-4),(3-5)}_{\substack{\text{next levels}\\ \text{are } n=4,\, n=5}} \rightarrow \underbrace{\mathbf{2,1,0,-1,-2}}_{m_l \text{ numbers}}$$

$$n=4 \rightarrow \underbrace{(4-1),(4-2),(4-3)}_{\substack{\text{previous levels are}\\ n=1,\, n=2,\, n=3}}; \quad \underbrace{(4-4)}_{\text{same level}} \quad \underbrace{(4-5),(4-6),(4-7)}_{\substack{\text{next levels are}\\ n=5,\, n=6,\, n=7}} \rightarrow \underbrace{\mathbf{3,2,1,0,-1,-2,-3}}_{m_l \text{ numbers}}$$

$$n=5 \rightarrow \underbrace{(5-1),(5-2),(5-3),(5-4)}_{\substack{\text{previous levels are } n=1,\\ n=2,\, n=3,\, n=4}} \quad \underbrace{(5-5)}_{\text{same level}} \quad \underbrace{(5-6),(5-7),(5-8),(5-9)}_{\substack{\text{next levels are } n=6,\\ n=7,\, n=8,\, n=9}} \rightarrow \underbrace{\mathbf{4,3,2,1,0,-1,-2,-3,-4}}_{m_l \text{ numbers}}$$

4. The Azimuthal number l .

The Azimuthal number l indicates the total number of electrons that can be added to a given energy level n comparing to its predecessor (actually meaning one additional sub shall to be populated at the next energy level). The electrons are set in pairs in sub shells formation dictated by their inverse orbital angular momentum (OAM) sign, their rotation is opposite in relation to each other which creates the attraction between them (or their inverse spins as the common presumption). The term l is expressed as; $l = n - 1$ (where n is the Principal number associated with the energy level) in sense of electrons allocation, it express the difference of their amount in relation to the previous energy level, which means that we need to subtract the total amount of electrons at the previous orbital level

That is indicated by $n = 1$ from the amount of electrons at the relevant level n. It yields the total number of electrons allocated for a new sub shell at the relevant level n!

For example:

The current principal number level ($n = 2$);

$$n = 2 \rightarrow \underbrace{\mathbf{4}(-\vec{\hbar}) \text{ and } \mathbf{4}(+\vec{\hbar})}_{\text{Total of } \mathbf{8} \text{ electrons}} \quad \text{According to sub section (1c)}$$

And previous principal number level ($n = 1$);

$$n = 1 \rightarrow \underbrace{\mathbf{1}(-\vec{\hbar}) \text{ and } \mathbf{1}(+\vec{\hbar})}_{\text{Total of } \mathbf{2} \text{ electrons}} \quad \text{According to sub section (1c)}$$

The subtraction of ($n = 1$) from ($n = 2$) yields the total number of electrons allocated for a new sub shell at the level $n = 2$ is:

$$\underbrace{\mathbf{3}(-\vec{\hbar}) \text{ and } \mathbf{3}(+\vec{\hbar})}_{\text{Total of } \mathbf{6} \text{ electrons}}$$

The Azimuthal number l at each level by using the previous example format process:

Note: The Magnetic quantum m_l values from section **3** are used here also to get a complete picture!

4a. The Aimuthal number l at $\mathbf{n = 1}$ (There are no sub shells or orbitals prior to $n = 1$)!

When ($l = 0$) it means no additional electrons to the ground state ($n = 1$) designated as sub shell **s**!

$$n = 1 \rightarrow \underbrace{\mathbf{1}(-\vec{\hbar}) \text{ and } \mathbf{1}(+\vec{\hbar})}_{\text{Total of } \mathbf{2} \text{ electrons}} \quad \text{According to sub section (1c)}$$

The electrons allocation in the new sub shell designated as **s** is as following:

$n = 1$, *subshell* **s**; $\mathbf{1}(-\vec{\hbar})$ and $\mathbf{1}(+\vec{\hbar}) \rightarrow$ consist of 1 pair of electrons in 1 *orbital* $\rightarrow$ **1,0, – 1**

4b. The Aimuthal number l at $\mathbf{n = 2}$

The current principal number level ($n = 2$);

$$n = 2 \rightarrow \underbrace{\mathbf{4}(-\vec{\hbar}) \text{ and } \mathbf{4}(+\vec{\hbar})}_{\text{Total of } \mathbf{8} \text{ electrons}} \quad \text{According to sub section (1c)}$$

And previous principal number level ($n = 1$);

$$n = 1 \rightarrow \underbrace{\mathbf{1}(-\vec{\hbar}) \text{ and } \mathbf{1}(+\vec{\hbar})}_{\text{Total of } \mathbf{2} \text{ electrons}} \quad \text{According to sub section (1c)}$$

The Azimuthal number l at $n = 2 \rightarrow (l = 2 - 1 = 1)$

The subtraction of ($n = 1$) from ($n = 2$) yields the total number of electrons allocated for a new sub shell designated as sub shell **p** at the level $n = 2$:

$$\underbrace{\mathbf{3}(-\vec{\hbar}) \text{ and } \mathbf{3}(+\vec{\hbar})}_{\text{Total of } \mathbf{6} \text{ electrons}}$$

The electrons allocation at $\mathbf{n = 2}$ with the new sub shell designated as **p** is as following:

$\mathbf{n} = 2$, *subshell* **s**; $\mathbf{1}(-\vec{\hbar})$ and $\mathbf{1}(+\vec{\hbar}) \rightarrow$ consist of 1 pair of electrons in 1 *orbital* $\rightarrow$ **1,0, – 1**

$n = 2$, *subshell* **p**; $\mathbf{3}(-\vec{\hbar})$ and $3(+\vec{\hbar}) \rightarrow$ consist of 3 pairs of electrons in 3 *orbitals* $\rightarrow$ **1,0, – 1**

4c. The Azimuthal number l at $\mathbf{n = 3}$

The current principal number level ($n = 3$);

$$n = 3 \rightarrow \underbrace{\mathbf{9}(-\vec{\hbar}) \text{ and } \mathbf{9}(+\vec{\hbar})}_{\text{Total of } \mathbf{18} \text{ electrons}} \quad \text{According to sub section (1c)}$$

And previous principal number level ($n = 2$);

$$n = 2 \rightarrow \underbrace{\mathbf{4}(-\vec{\hbar}) \text{ and } \mathbf{4}(+\vec{\hbar})}_{\text{Total of } \mathbf{8} \text{ electrons}} \quad \text{According to sub section (1c)}$$

The Azimuthal number l at $n = 3 \rightarrow (l = 3 - 1 = 2)$

The subtraction of ($n = 2$) from ($n = 3$) yields the total number of electrons added to a new sub shell designated as **d** at the level $n = 3$:

$$\underbrace{\mathbf{5}(-\vec{\hbar}) \text{ and } \mathbf{5}(+\vec{\hbar})}_{\text{Total of } \mathbf{10} \text{ electrons}}$$

The electrons allocation at $\mathbf{n = 3}$ with the new sub shell designated as **d** is as following:

$n = 3$, *subshell* **s**; $\mathbf{1}(-\vec{\hbar})$ and $\mathbf{1}(+\vec{\hbar}) \rightarrow$ consist of 1 pairs of electron in 1 *orbitals.* $\rightarrow$ **1,0, − 1**

$n = 3$, *subshell* **p**; $\mathbf{3}(-\vec{\hbar})$ and $\mathbf{3}(+\vec{\hbar}) \rightarrow$ consist of 3 pairs of electrons in 3 *orbitals.* $\rightarrow$ **1,0, − 1**

$n = 3$, *subshell* **d**; $\mathbf{5}(-\vec{\hbar})$ and $\mathbf{5}(+\vec{\hbar}) \rightarrow$ consist of 5 pairs of electrons in 5 *orbitals* $\rightarrow$ **2,1,0, − 1,−2**

4d. The Azimuthal number l at $\mathbf{n = 4}$

The current principal number level ($n = 4$);

$$n = 4 \rightarrow \underbrace{\mathbf{16}(-\hbar) \text{ and } \mathbf{16}(+\hbar)}_{\text{Total of } \mathbf{32} \text{ electrons}} \text{ According to sub section (1c)}$$

And previous principal number level ($n = 3$);

$$n = 3 \rightarrow \underbrace{\mathbf{9}(-\vec{\hbar}) \text{ and } \mathbf{9}(+\vec{\hbar})}_{\text{Total of } \mathbf{18} \text{ electrons}} \text{ According to sub section (1c)}$$

The Azimuthal number l at $n = 4 \rightarrow (l = 4 - 1 = 3)$

The subtraction of ($n = 3$) from ($n = 4$) yields the total number of electrons added to a new sub shell designated as **f** at the level $n = 4$:

$$\underbrace{\mathbf{7}(-\vec{\hbar}) \text{ and } \mathbf{7}(+\vec{\hbar})}_{\text{Total of } \mathbf{14} \text{ electrons}}$$

The electrons allocation at $\mathbf{n = 4}$ with the new sub shell designated as **f** is as following:

$n = 4$, *subshell* **s**; $\mathbf{1}(-\vec{\hbar})$ and $\mathbf{1}(+\vec{\hbar}) \rightarrow$ consist of 1 pair of electrons in 1 *orbital* $\rightarrow$ **1,0, − 1**

$n = 4$, *subshell* **p**; $\mathbf{3}(-\vec{\hbar})$ and $\mathbf{3}(+\vec{\hbar}) \rightarrow$ consist of 3 pairs of electrons in 3 *orbitals.* $\rightarrow$ **1,0, − 1**

$n = 4$, *subshell* **d**; $\mathbf{5}(-\vec{\hbar})$ and $\mathbf{5}(+\vec{\hbar}) \rightarrow$ consist of 5 pairs of electrons in 5 *orbitals* $\rightarrow$ **2,1,0, − 1,−2**

$n = 4$, *subshell* **f**; $\mathbf{7}(-\vec{\hbar})$ and $\mathbf{7}(+\vec{\hbar}) \rightarrow$ consist of 7 pairs of electrons in 7 *orbitals* $\rightarrow$ **3,2,1,0, − 1,−2,−3**

4e. The Azimuthal number l at $n = 5 \rightarrow (l = 4 - 1 = 3)$

The current principal number level ($n = 5$) contains;

$$n = 5 \rightarrow \underbrace{\mathbf{25}(-\hbar) \text{ and } \mathbf{25}(+\hbar)}_{\text{Total of } \mathbf{50} \text{ electrons}} \text{ According to sub section (1c)}$$

And previous principal number level ($n = 3$) contains;

$$n = 4 \rightarrow \underbrace{\mathbf{16}(-\hbar) \text{ and } \mathbf{16}(+\hbar)}_{\text{Total of } \mathbf{32} \text{ electrons}} \text{ According to sub section (1c)}$$

The subtraction of ($n = 2$) from ($n = 3$) yields the total number of electrons added to a new sub shell designated as **g** at the level $n = 5$:

$$\underbrace{\mathbf{9}(-\vec{\hbar}) \text{ and } \mathbf{9}(+\vec{\hbar})}_{\text{Total of } \mathbf{18} \text{ electrons}}$$

The electrons allocation at $\mathbf{n = 5}$ with the new sub shell designated as **g** is as following:

$n = 5$, *subshell* **s**; $\mathbf{1}(-\vec{\hbar})$ and $\mathbf{1}(+\vec{\hbar}) \rightarrow$ consist of 1 pair of electrons in 1 *orbital* $\rightarrow$ **1,0, − 1**

$n = 5$, *subshell* **p**; $\mathbf{3}(-\vec{\hbar})$ and $\mathbf{3}(+\vec{\hbar}) \rightarrow$ consist of 3 pairs of electrons in 3 *orbitals.* $\rightarrow$ **1,0, − 1**

$n = 5$, *subshell* **d**; $\mathbf{5}(-\vec{\hbar})$ and $\mathbf{5}(+\vec{\hbar}) \rightarrow$ consist of 5 pairs of electrons in 5 *orbitals* $\rightarrow$ **2,1,0, − 1,−2**

$n = 5$, *subshell* **f**; $\mathbf{7}(-\vec{\hbar})$ and $\mathbf{7}(+\vec{\hbar}) \rightarrow$ consist of 7 pairs of electrons in 7 *orbitals* $\rightarrow$ **3,2,1,0, − 1,−2,−3**

$n = 5$, *subshell* **g**; $\mathbf{9}(-\vec{\hbar})$ and $\mathbf{9}(+\vec{\hbar}) \rightarrow$ consist of 9 pairs of electrons in 9 *orbitals* $\rightarrow$ **4,3,2,1,0,−1,−2,−3,−4**

5. The Electron's and the other Fermions Spins**:**
I copy here the equations related to $\vec{L}$ from sector **2** in this chapter with numeric values for each.

$$n = 2 \rightarrow \underbrace{l = (2-1)}_{\text{previous level is } n=1} \rightarrow l = 1 \rightarrow \vec{L} = \pm\sqrt{1 \times 2} \times \vec{\hbar}; \quad \vec{L} = \sqrt{2} \times \vec{\hbar} = \pm 1.4142\vec{\hbar} \rightarrow \vec{L} \cong \pm 1.5 \times \vec{\hbar}$$

$$n = 3 \rightarrow \underbrace{l = (3-1)}_{\text{previous level is } n=2} \rightarrow l = 2 \rightarrow \vec{L} = \pm\sqrt{2 \times 3} \times \vec{\hbar}; \quad \vec{L} = \pm\sqrt{6} \times \vec{\hbar} = 2.4494\vec{\hbar} \rightarrow \vec{L} \cong \pm 2.5 \times \vec{\hbar}$$

$$n = 4 \rightarrow \underbrace{l = (4-1)}_{\text{previous level is } n=3} \rightarrow l = 3 \rightarrow \vec{L} = \pm\sqrt{3 \times 4} \times \vec{\hbar}; \quad \vec{L} = \pm\sqrt{12} \times \vec{\hbar} = 3.4641\vec{\hbar} \rightarrow \vec{L} \cong \pm 3.5 \times \vec{\hbar}$$

From viewing at the results of $\vec{L}$, we can observe that the orbital quantum number operator <u>points to or from a middle distance between two close **n** levels</u>**:- (1.5),(2.5),(3.5),(4.5),(5.5)**
It seemingly creates a kind of levels for **n** (between those expressed by integers) but with half integer value for **n**, and it should look as: $n = 0.5, 1, 1.5, 2, 2.5, 3, 3.5, 4, 4.5, 5, 5.5...$
It suggests that the spin of the electron from the start is an integral component of $\vec{L}$ the angular momentum and both reflects the 'spin-orbit coupling' effect in reality, although it cannot be measured directly. We are measuring their projections $\vec{L}_z = \mathbf{m}_l\vec{\hbar}$ and $\vec{s}_z = \pm\mathbf{0.5} \times \vec{\hbar}$ along the **z** axis and sum those up to get the solution for 'spin-orbit coupling' effect in reality. Following the suggestion, means that $\vec{L}$ is already the 'spin-orbit coupling' fulfilling the role of $\vec{J}$, meaning that $\vec{L} = \vec{L}_z + \vec{s}_z$ **.**

$$n = 2 \rightarrow \vec{L} \cong \pm 1.5 \times \vec{\hbar} \text{ equals} \rightarrow \underbrace{\vec{L}_z = \pm 1 \times \vec{\hbar}}_{\text{projected } \vec{L}} \text{ plus } \underbrace{\vec{s}_z = \pm 0.5 \times \vec{\hbar}}_{\text{projected spin } \vec{s}}$$

$$n = 3 \rightarrow \vec{L} \cong \pm 2.5 \times \vec{\hbar} \text{ equals} \rightarrow \underbrace{\vec{L}_z = \pm 2 \times \vec{\hbar}}_{\text{projected } \vec{L}} \text{ plus } \underbrace{\vec{s}_z = \pm 0.5 \times \vec{\hbar}}_{\text{projected spin } \vec{s}}$$

$$n = 4 \rightarrow \vec{L} \cong \pm 3.5 \times \vec{\hbar} \text{ equals} \rightarrow \underbrace{\vec{L}_z = \pm 3 \times \vec{\hbar}}_{\text{projected } \vec{L}} \text{ plus } \underbrace{\vec{s}_z = \pm 0.5 \times \vec{\hbar}}_{\text{projected spin } \vec{s}}$$

Note: The presence of the electron's spin within $\vec{L}$ (constitute one entity with shared spatial orientation) <u>provides a new interpretation for the meaning of the existence of the higher half spin values</u> $1.5\vec{\hbar}, 2.5\vec{\hbar}, 3.5\vec{\hbar}, 4.5\vec{\hbar}....$, which the quantum theory predicts for the elementary fermions which they do not seem to exist. I do not compare this with the case of the 'composite Baryons'.

Here are the spins per orbitals according to: - $\underbrace{\vec{L}_z = \mathbf{m}_l\vec{\hbar}}_{\text{projected } \vec{L}} \text{ plus } \underbrace{\vec{s}_z = \pm 0.5 \times \vec{\hbar}}_{\text{projected spin } \vec{s}}$ $\mathbf{m}_l$ from sector 3 this chapter. Zero value express an opposed spins vectors on plane x-y that are perpendicular to z axis!

$$\underbrace{n = 1}_{\text{For } \vec{L} = 0.5\vec{\hbar}} \rightarrow 0.5\vec{\hbar}, 0, -0.5\vec{\hbar}$$

$$\underbrace{n = 2}_{\text{For } \vec{L} = 1.5\vec{\hbar}} \rightarrow 1.5\vec{\hbar}, 0, -1.5\vec{\hbar}$$

$$\underbrace{n = 3}_{\text{For } \vec{L} = 2.5\vec{\hbar}} \rightarrow 2.5\vec{\hbar}, 1.5\vec{\hbar}, 0, -1.5\vec{\hbar}, -2.5\vec{\hbar}$$

$$\underbrace{n = 4}_{\text{For } \vec{L} = 3.5\vec{\hbar}} \rightarrow 3.5\vec{\hbar}, 2.5\vec{\hbar}, 1.5\vec{\hbar}, 0, -1.5\vec{\hbar}, -2.5\vec{\hbar}, -3.5\vec{\hbar}$$

* The puzzle of the proton's charge radius is solved by this theory!

This section deals with the discrepancy found in the charge radius of the proton measured at different institutes' laboratories. The discrepancy exist between the last recommend value $8.768 \times 10^{-16}[m]$ of the International Council for Science's Committee on Data for Science & Technology (CODATA) set on in 2006, and a more recent result (2010) corresponding with the experimental findings of Randolf Pohl from Max Planck Institute of Quantum Optics and his colleagues, that set the charge proton radius on $8.4184 \times 10^{-16}[m]$ in a muonic Hydrogen experiment. This section provides a theoretical explanation for this discrepancy and introduces a finite theoretical result $r_{pcr} = 8.4124 \times 10^{-16}[m]$ that corresponds with the experimental findings of Randolf Pohl and his colleagues.

The external high-energy electrons that are fired and penetrate the proton in a sub atomic accelerator, are interacting with the two *up* quarks at the second orbital level within the proton and their ultimate approaching distance within the proton is marked by the so called "charge radius". This radius presents only a part of its actual radius which was assumed to be the entire radius!

The electron "treats" the proton's orbital radiuses as an extension of the orbitals in the Hydrogen atom, a kind of a new ground state for it while it penetrates the proton. So for the orbital levels radius within the proton in terms of the Hydrogen orbitals we write:

$$r_{\text{n}} = \text{n}^2 \times r_p\sqrt{1-\beta^2}/\text{n}$$

Using Eq. 242 with the last term, forms:

$$\vec{L}_{q(\text{n})} = m_q \times v_{q(\text{n})} \times (\text{n}^2 \times r_p\sqrt{1-\beta^2}/\text{n})$$

Substituting for **n = 2** (*up* quarks level) $m_q = 1/3\, m_p, v_{q(\text{n}=2)} = c(2\beta)$ and $r_{(\text{n}=2)} = r_p\sqrt{1-\beta^2}/2$ in yields:

$$\vec{L}_{q(\text{n}=2)} = 1/3\, m_p \times c(2\beta) \times 2^2 \times r_p\sqrt{1-\beta^2}/2 \text{ or} \rightarrow \vec{L}_{q(\text{n}=2)} = 1/3\, m_p \times c \times [\beta \times 2^2 \times r_p\sqrt{1-\beta^2}]$$

The term in the square brackets is the Proton's Charge Radius marked as r_{pcr}:

$$r_{pcr} = 4 \times \beta \times r_p \times \sqrt{1-\beta^2}$$

Substituting:- $\beta = 0.178145016$, $r_p = 1.199746 \times 10^{-15}[m]$ and $\sqrt{1-\beta^2} = 0.984004245$ yields:

$$r_{pcr} = 4 \times 0.178145016 \times 1.19974618 \times 10^{-15}[m] \times 0.984004245 = 8.4124 \times 10^{-16}[m]$$

This result matches the experimental finding of Randolf Pohl from the Max Planck Institute of Quantum Optics and his colleagues, who revealed a discrepancy between their findings on proton's charge radius $8.4184 \times 10^{-16}[m]$ in a Muonic Hydrogen experiment (kind of an exotic atom), and the International Council for Science's Committee on Data for Science & Technology (CODATA) recommended value of $8.768 \times 10^{-16}[m]$ settled on in 2006, after reviewing results from experiments that involves scattering of electrons off protons and hydrogen spectroscopy method. The experiments conducted on finding the charge radius of the proton are using either electrons that are the obvious choice or a muon that create short-lived Muonic Hydrogen. The electrons and muons that penetrate the proton in a collision are actually exposing the maximum distance that they can reach the second orbital within the proton occupied by the two *up* quarks, but in fact the real proton radius is set by the *down* quark at the first Orbital within the proton with a relativistic value r_p':

$$r_p' = r_p\sqrt{1-\beta^2} \cong 1.1808 \times 10^{-15}[m]$$

An article that discusses this issue is the following.

Muonic Hydrogen and the Proton Radius Puzzle. Randolf Pohl et al. in *Annual Review of Nuclear and Particle Science*, Vol. 63, pages 175-204; October 2013.

13

The Energy Levels within the Proton and the Neutron, And the origin of the Weak Force $W^{\pm}$ Boson

In the previous chapter I have described the proton and the neutron structures as incorporating three energy levels within on which the quarks are orbiting. The *down* and *up* quarks in the proton and the neutron occupy the first and the second energy levels. The significance of the third energy level will be clarified later in this chapter on the discussion of the origin of the weak force $W^{\pm}$ boson!

Let us now find the three energy levels within the neutron and the proton.

Note: The energy levels are almost identical for both the neutron and the proton!

To obtain the kinetic energy for each orbital level within the proton or the neutron we perform the following process. We start by using Eq. 18 for the proton:

$$e^2 = \frac{m_p v_p^2 4\pi r_p^2}{6}\left(\frac{m_p}{m_e}\right) \quad \text{Or} \rightarrow \quad e^2 = \frac{1}{2} m_p v_p^2 \times 4\pi r_p^2 \times \frac{1}{3}\left(\frac{m_p}{m_e}\right)$$

Subtract both flanks by $4\pi r_p^2$, and substitute $v_p = c \times \beta$ and $\hbar = m_p v_p r_p$, it yields:

$$\frac{e^2}{4\pi r_p^2} = \frac{1}{2}\hbar c\beta \times \frac{4\pi r_p}{4\pi r_p^2} \times \frac{1}{3}\left(\frac{m_p}{m_e}\right) \quad \text{Or} \rightarrow \quad \frac{e^2}{4\pi r_p^2} = \frac{\hbar \times c \times \beta}{2 \times r_p} \times \frac{1}{3}\left(\frac{m_p}{m_e}\right)$$

The right flank of the last equation represents the general term of E_n the kinetic energy of the orbital level on which the relevant quark revolves. To describe the discrete kinetic energy of each **n** (orbital level), we need to substitute the following in the general term: Substitute $(n\beta)$ instead of β, $\left(\frac{r_p}{n}\right)$ instead of r_p, replace $\hbar$ with $\vec{L}_n$ and $n = 1,2,3$ (the number of the relevant level for the three inner levels of the baryon), it forms:

(250)

For the proton $$E_n = \frac{\vec{L}_n \times c \times (n\beta)}{2 \times \frac{r_p}{n}} \times \frac{1}{3}\frac{m_p}{m_e}$$

For the neutron $$E_n = \frac{\vec{L}_n \times c \times (n\beta)}{2 \times \frac{r_n}{n}} \times \frac{1}{3}\frac{m_n}{m_e}$$

Where:

E_n -is the kinetic energy for each orbital level within the proton or the neutron.

r_n -is the relevant orbital radius within the proton or the neutron on which the quark is orbiting.

v_n -is the quark's orbital velocity: $v_n = c \times (n\beta)$

$\vec{L}_n$ -is representing the quark's orbital angular momentum at the relevant orbital level.

n -is the orbital number: n = 1, 2, 3.

$\hbar$ - is the reduced Planck constant - $\left(\frac{h}{2\pi}\right)$.

Notes:

1. Because there are three orbitals we assume that each of the three quarks within the neutron or the proton share one-third of the proton or the neutron mass apiece.

   For the neutron the quark mass is $1/3\, m_n$ and for the proton the quark mass is $1/3\, m_p$ so we write

   the baryon's quark mass to electron mass ratio: $\dfrac{1}{3}\dfrac{m_n}{m_e}$ and $\dfrac{1}{3}\dfrac{m_p}{m_e}$ respectively.

2. For the neutron I am presenting the energy level at the third orbital (for the other levels it should be identical as for the proton that comes next) for its importance.

   The third orbital plays a main role in the decaying process thru the weak force as it will be shown later in the next chapters!

a. The energy level related to the third orbital within the neutron:

The theoretical value of the neutron mass to electron mass ratio from Eq. 167 is:

$$\frac{m_n}{m_e} = 1838.682747$$

At the third level n=3: this is the lowest inner orbital in the neutron.

Substituting the following data values into Eq. 250 for the neutron

$$\frac{m_{neutron\ quark}}{m_e} = \frac{1}{3}\frac{m_n}{m_e} = \frac{1}{3} \times 1838.682747$$

The theoretical value of the non-relativistic neutron radius r_n from Eq. 169 is:

$$r_n = 1.19810139 \times 10^{-15}\,[m]$$

The theoretical value of $\vec{L}_{quark}$ at the third orbital level according to Eq. 248 is:

$$\vec{L}_{n=3} = \hbar = 1.054585924 \times 10^{-34}\,[J\ s]$$

Note: The quark's orbital angular momentum $\vec{L}_n$ is a factor of one third or two thirds of $\hbar$ as I showed in Eq. 244 and Eq. 246!

This means that for the first level it has $1/3\hbar$ for the second level $2/3\,\hbar$ and for the third level $\hbar$.

In addition the $\vec{L}_n$ can be either negative or positive depending on the quark's orbiting direction at the relevant orbital. For a particle with quarks at the first orbital level $\vec{L}_n$ is negative, for the second $\vec{L}_n$ is positive and for the third $\vec{L}_n$ is negative again. For an antiparticle with anti quarks at the first orbital level $\vec{L}_n$ is positive, for the second $\vec{L}_n$ is negative and for the third $\vec{L}_n$ is positive again. In either way the total $\vec{L}_n$ of the baryon or the anti baryon can be zero or $\hbar$. The orbiting direction is usually indicated by an arrow, for instance negative is marked ($\hbar\downarrow$) and positive is marked ($\hbar\uparrow$). Please notice that the **total Baryon spin** known as j^p (positive or negative) will be the sum of $\vec{L}_n$ and the contribution of each quark's unique spin. So the total Baryon spin can be a fraction like $1/2\,\hbar$ or $3/2\,\hbar$ and so forth.

The theoretical value for the speed of light in vacuum c as was calculated by Eq. 96 is:

$$c = 2.99792344 \times 10^8\,[m\ s^{-1}]$$

The key constant: $\beta = 0.178145016$

Substituting all the data values presented so far into Eq. 250 yields the theoretical value of the kinetic energy at the third orbital level within the neutron, yields:

$$(251)\quad E_{n=3} = \frac{\vec{L}_{n=3} \times c \times (3\beta)}{2 \times \dfrac{r_n}{3}} \times \frac{1}{3}\frac{m_n}{m_e} = 1.296524335 \times 10^{-8}\,[J] = 80.92 \times 10^9\,[ev] = 80.92\,[Gev]$$

And by dividing the last result by c^2 (the square speed of light in vacuum) we receive the equivalent mass of this energy:

$$1.296524335 \times 10^{-8}[J] \times \frac{1}{\left(2.99792344 \times 10^{8}\right)^{2}\left[m\, s^{-1}\right]^{2}} = 1.44257897 \times 10^{-25}[kg]$$

The last result: $1.44257897 \times 10^{-25}[kg]$ is equivalent to nearly **86 protons**!

Note: The result of Eq. 251 points on the origin of the W^- boson that has a mass of $\sim 80[Gev]$ and it predicts a position at the third level for a quark or more that their total charge sum up to e^- in the three energy levels model of the neutron and proton, and apparently of all baryons.

The equivalent mass of the energy levels are not detected while the quark or quarks are orbiting within the baryons, because the quarks doesn't radiate as long as they are orbiting at the energy levels, therefore the kinetic energy $\sim 80[Gev]$ related to the third level for instance is not detected. It is detected only when the quark or quarks are being emitted out along with a γ photon that carries the released kinetic energy $\sim 80[Gev]$ from the third energy level and jointly they form the W^- boson!

In general: it seems that particles that consist of $W^{\pm}$ boson which involves in their weak decay process, emits a quark or more from the third level along with a γ photon respectively!

b. The energy levels related to the three orbitals within the proton:

At the third level n=3: this is the lowest inner orbital in the proton.

Substituting the following data values into Eq. 250 for the proton

$$\frac{m_{proton\ quark}}{m_e} = \frac{1}{3}\frac{m_p}{m_e} = \frac{1}{3} \times 1836.151064$$

$$r_p = 1.199746186 \times 10^{-15}[m]$$

$$\vec{L}_{n=3} = \hbar = 1.054585924 \times 10^{-34}[J\ s]$$

$$c = 2.99792344 \times 10^{8}[m\, s^{-1}]$$

$$\beta = 0.178145016$$

Yield the theoretical value of the kinetic energy at the third orbital level within the proton.

$$(252) \qquad E_{n=3} = \frac{\vec{L}_{n=3} \times c \times (3\beta)}{2 \times \frac{r_p}{3}} \times \frac{1}{3}\frac{m_p}{m_e} = 1.292964123 \times 10^{-8}[J] = 80.7 \times 10^{9}[ev] = 80.7[Gev]$$

And by dividing the result in Eq. 252 by c^2 (the square speed of light in vacuum) we receive the equivalent mass of this energy:

$$1.292964123 \times 10^{-8}[J] \times \frac{1}{\left(2.99792344 \times 10^{8}\right)^{2}\left[m\, s^{-1}\right]^{2}} = 1.438617698 \times 10^{-25}[kg]$$

The last result: $1.438617698 \times 10^{-25}[kg]$ is equivalent to nearly **86 protons**!

At the first level n=1: this is the outer orbital which creates the surface of the proton.

Substituting the following data values into Eq. 250 for the proton

$$\frac{m_{proton\ quark}}{m_e} = \frac{1}{3}\frac{m_p}{m_e} = \frac{1}{3} \times 1836.151064$$

$$r_p = 1.199746186 \times 10^{-15}[m]$$

$$\vec{L}_{n=1} = 1/3\, \hbar = 3.515286413 \times 10^{-35}[J\ s]$$

$$c = 2.99792344 \times 10^{8}[m\, s^{-1}]$$

$\beta = 0.178145016$

Yield the theoretical value of the kinetic energy of the quark at the first orbital level.

(253)

$$E_{n=1} = \frac{\vec{L}_{n=1} \times c \times (\beta)}{2 \times r_p} \times \frac{1}{3}\frac{m_p}{m_e} = 4.788756 \times 10^{-10}[J] = 2.989 \times 10^{9}[ev] = 2.989[Gev]$$

And by dividing the result in Eq. 253 by c^2 (the square speed of light in vacuum) we receive the equivalent mass of this energy:

$$4.788756 \times 10^{-10}[J] \times \frac{1}{\left(2.99792344 \times 10^{8}\right)^{2}\left[m\,s^{-1}\right]^{2}} = 5.3282137 \times 10^{-27}[kg]$$

The last result: $5.3282137 \times 10^{-27}[kg]$ is equivalent to nearly **3 protons**!

At the second level n=2: this is the mid inner orbital in the proton.

Substituting the following data values into Eq. 250 for the proton

$$\frac{m_{proton\ quark}}{m_e} = \frac{1}{3}\frac{m_p}{m_e} = \frac{1}{3} \times 1836.151064$$

$$r_p = 1.199746186 \times 10^{-15}[m]$$

$$\vec{L}_{n=2} = 2/3\,\hbar = 7.030572824 \times 10^{-35}[J\,s]$$

$$c = 2.99792344 \times 10^{8}[m\,s^{-1}]$$

$$\beta = 0.178145016$$

Yield the theoretical value of the quark's kinetic energy at the second orbital level.

$$(254)\quad E_{n=2} = \frac{\vec{L}_{n=2} \times c \times (2\beta)}{2 \times \frac{r_p}{2}} \times \frac{1}{3}\frac{m_p}{m_e} = 3.831 \times 10^{-9}[J] = 2.3911 \times 10^{10}[ev] = 23.911[Gev]$$

And dividing the result in Eq. 254 by c^2 (the square of speed of light in vacuum) we receive the equivalent mass of this energy.

$$3.831 \times 10^{-9}[J] \times \frac{1}{\left(2.99792344 \times 10^{8}\right)^{2}\left[m\,s^{-1}\right]^{2}} = 4.262565 \times 10^{-26}[kg]$$

The last result: $4.262565 \times 10^{-26}[kg]$ is equivalent to nearly **26 protons**!

In general: During the subatomic particles interactions, the quarks positioned at the lower energy levels passes through excited state modes reflected by shifting to a higher energy levels that changes their characteristics according to their new position. A quark or quarks that reaches the third level in the process, triggers the weak force decay process while being emitted out along with a γ photon that carries the released kinetic energy $\sim 80[Gev]$ from the third energy level and jointly they form the massive $W^{\pm}$ boson that in addition to other interaction products creates a transformation of the original particle in question. For instance let's see the following scenario.

Scenario: A proton and anti proton collides in a particles accelerator involving an interaction with the second and the third energy levels within the proton.

Note: Because there is symmetry between the outcomes, the collision between the proton and the anti proton generates similar two new particles, emitted in opposed directions. I will describe the event of the proton's decay in the collision and please remember that this process occurs simultaneously with the antiproton as well. After the proton has collided, the single *down* quark $\left(-\frac{1}{3}e^{-}\right)$ at the first level absorbs enough energy from the collision to elevate to the third level changing its charge to e^{-}

and one of the *up* quarks $\left(+\frac{2}{3}e^{+}\right)$ at the second level absorbs enough energy from the collision to elevate to the third level as well changing its charge to e^{+}. The other *up* quark $\left(+\frac{2}{3}e^{+}\right)$ at the second level remains intact.

This new particle generated in the collision (in excited state mode) has a mass that is equivalent to the total energy absorbed in the process according to the following calculations outcome:

At the **third level n=3** in the proton, the energy level is according to Eq. 252:

$$E_{n=3} = \frac{\vec{L}_{n=3} \times c \times (3\beta)}{2 \times \frac{r_p}{3}} \times \frac{1}{3}\frac{m_p}{m_e} = 1.292964123 \times 10^{-8}[J] = 80.7 \times 10^{9}[ev] = 80.7[Gev]$$

At the **second orbital level n=2** in the proton, the energy level is according to Eq. 254:

$$E_{n=2} = \frac{\vec{L}_{n=2} \times c \times (2\beta)}{2 \times \frac{r_p}{2}} \times \frac{1}{3}\frac{m_p}{m_e} = 3.831 \times 10^{-9}[J] = 2.3911 \times 10^{10}[ev] = 23.911[Gev]$$

So the new massive particle that was generated in the collision and consists of two mesons at the third level and one up quark at the second level has a mass that is equivalent to the total energy absorbed:

$$E_{(\text{total energy})} = 2 \times 80.7[Gev] + 23.911[Gev] = 185.31[Gev]$$

This massive particle generated in the collision is the **top quark** which according to the experiments has a mass equivalent to $175[Gev]$ that is emitted out along with a γ photon released as a **transverse energy** $\gamma = 10.31[Gev]$ which is equivalent to what is left out from the total of $185.31[Gev]$ energy!

The **top quark** consist of three inner particles, two of them are Pi mesons at the third level and neutralizes each other due to their opposite electric charges e^{-} and e^{+} that prevents their detection, thus seemingly it possess a fraction of the elementary charge observed in experiments $\left(+\frac{2}{3}e^{+}\right)$ which is actually the charge of the third quark (the *up* quark positioned at the second level and is not neutralized!).

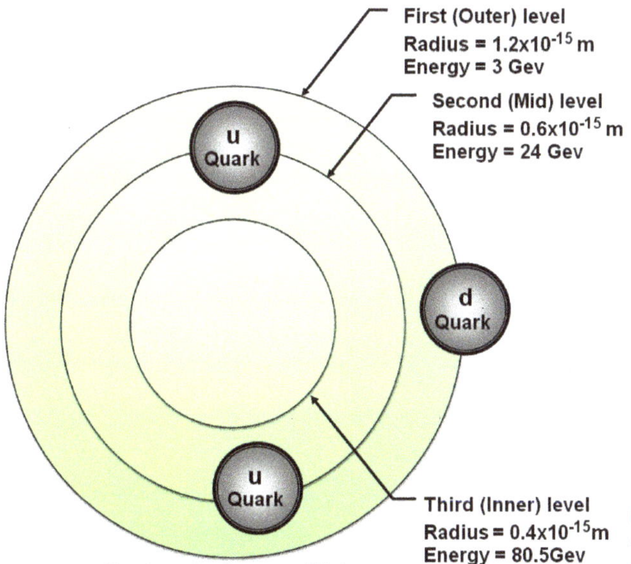

Proton before collision

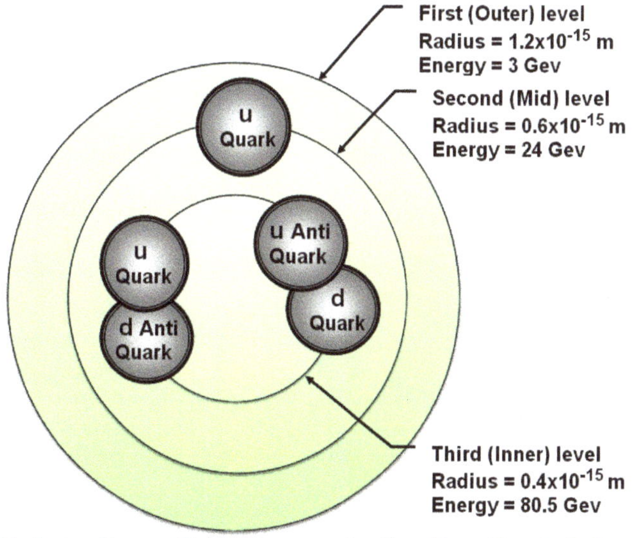

Proton after collision represents the Top Quark before decaying. One u Quark shifts to third level, transformed into a positive π meson thru a gluon fusion. The d Quark shifts to third level, transformed into a negative π meson thru a gluon fusion. The second u Quark remains intact at the second level.

The final process can be presented by the following schema here.

Jet consist of *b quark* thru gluons fusion $b \rightarrow \underbrace{u\bar{u}d}_{s}\underbrace{u\bar{u}d}_{s}\underbrace{d\bar{d}\bar{d}}_{\bar{s}}\ or\ \underbrace{u\bar{u}u}_{c}+\underbrace{\bar{u}d\bar{d}}_{\bar{c}}+\underbrace{dd\bar{d}}_{s}$

(On Gluon fusion see page 152)

The Jet consist of two Strange quarks and one Strange anti quark
or a Charm quark and a Charm anti quark and one Strange quark
or Three Pi mesons and one Strange quark. Please see new interpretation
for the second and third quarks generation at chapter 15!

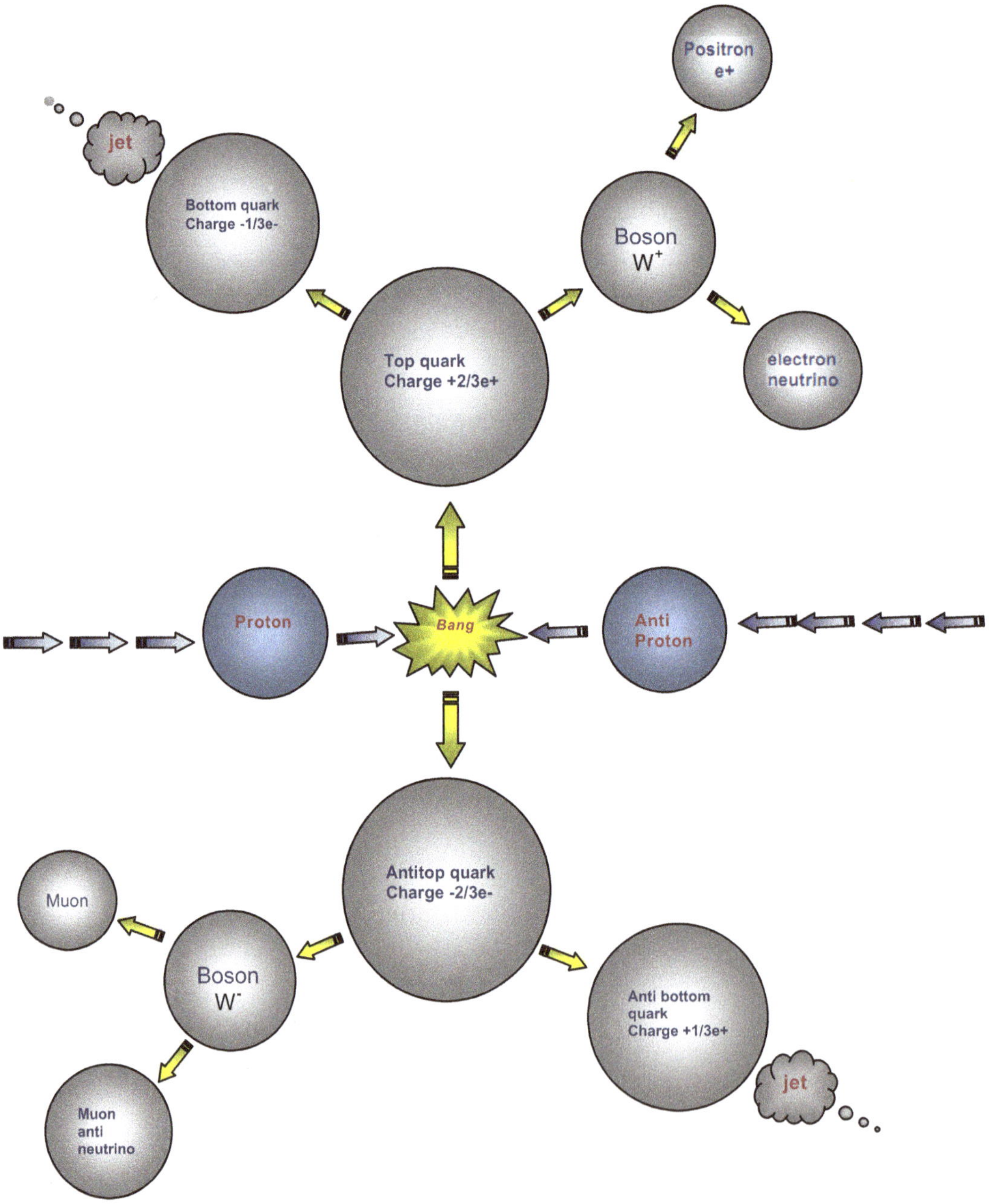

A schema shows the discovery of the Top quark, while a proton and antiproton collides in a Particle Accelerator.

* The weak force decay thru $W^{\pm}$ boson mechanism.

a. The neutron excited state: The neutron's inner structure prior to its decaying through the weak force is as following: While exchanging pions within the nucleolus, the neutron absorbs a neutral meson π° that changes its inner binding energy configuration by a new generated π^{-} meson (see the schema below) that reaches the third orbital level in a momentarily captured status which creates an **excited neutron** with a neutral charge and $1/2\hbar$ spin. The π^{-} meson doesn't radiate while orbiting at the third level and therefore the kinetic energy ~80[Gev] related to the third level is not noticeable!

neutron

spin		orbital	charge
$+\frac{1}{2}\hbar\uparrow$	u	$\uparrow +\frac{2}{3}\hbar$	$+\frac{2}{3}e^{+}$
$-\frac{1}{2}\hbar\downarrow$	d	$\downarrow -\frac{1}{3}\hbar$	$-\frac{1}{3}e^{-}$
$+\frac{1}{2}\hbar\uparrow$	d	$\downarrow -\frac{1}{3}\hbar$	$-\frac{1}{3}e^{-}$

Total spin : $+1/2\hbar$
Total orbital : 0
Total spin + orbital : $+1/2\hbar$
neutral charge :

$+$

π°

spin		orbital	charge
$-\frac{1}{2}\hbar\downarrow$	$\bar{u}$	$\downarrow -\frac{2}{3}\hbar$	$-\frac{2}{3}e^{-}$
$+\frac{1}{2}\hbar\uparrow$	u	$\uparrow +\frac{2}{3}\hbar$	$+\frac{2}{3}e^{+}$

Total spin : 0
Total orbital : 0
Total spin + orbital : 0
neutral charge :

$\rightarrow$

excited neutron

spin		orbital	charge
$+\frac{1}{2}\hbar\uparrow$	u	$\uparrow +\frac{2}{3}\hbar$	$+\frac{2}{3}e^{+}$
$+\frac{1}{2}\hbar\uparrow$	u	$\uparrow +\frac{2}{3}\hbar$	$+\frac{2}{3}e^{+}$
$-\frac{1}{2}\hbar\downarrow$	d	$\downarrow -\frac{1}{3}\hbar$	$-\frac{1}{3}e^{-}$

Total spin : $+1/2\hbar$
Total orbital : $+\hbar$
Total spin + orbital : $+3/2\hbar$
Charge : e^{+}

π^{-} positioned at the third level of the excited neutron

spin		orbital	charge
$-\frac{1}{2}\hbar\downarrow$	$\bar{u}$	$\downarrow -\frac{2}{3}\hbar$	$-\frac{2}{3}e^{-}$
$+\frac{1}{2}\hbar\uparrow$	d	$\downarrow -\frac{1}{3}\hbar$	$-\frac{1}{3}e^{-}$

Total spin : 0
Total orbital : $-\hbar$
Total spin + orbital : $-\hbar$
Charge : e^{-}

Total j^{p}(spin + orbital) = $+1/2\hbar$
Neutral charge

The **excited neutron**

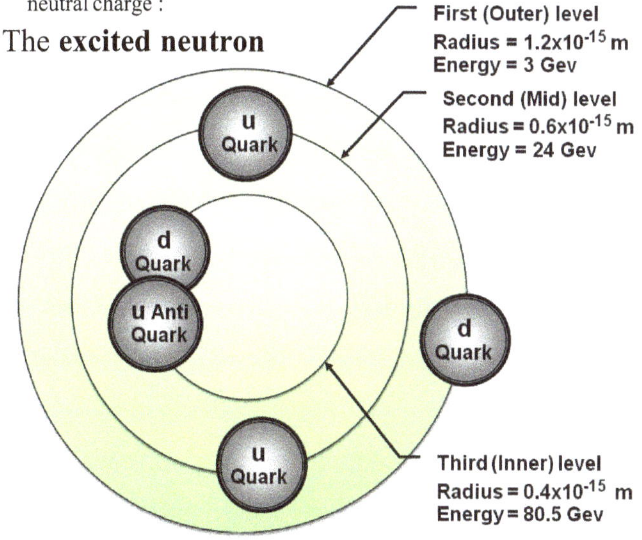

Excited Neutron before the negative π meson at the third energy level is being emitted out and transformed into a Proton.

excited neutron

spin		orbital	charge
$+\frac{1}{2}\hbar\uparrow$	u	$\uparrow +\frac{2}{3}\hbar$	$+\frac{2}{3}e^{+}$
$+\frac{1}{2}\hbar\uparrow$	u	$\uparrow +\frac{2}{3}\hbar$	$+\frac{2}{3}e^{+}$
$-\frac{1}{2}\hbar\downarrow$	d	$\downarrow -\frac{1}{3}\hbar$	$-\frac{1}{3}e^{-}$

Total spin : $+1/2\hbar$
Total orbital : $+\hbar$
Total spin + orbital : $+3/2\hbar$
Charge : e^{+}

π^{-} positioned at the third level of the excited neutron

spin		orbital	charge
$-\frac{1}{2}\hbar\downarrow$	$\bar{u}$	$\downarrow -\frac{2}{3}\hbar$	$-\frac{2}{3}e^{-}$
$+\frac{1}{2}\hbar\uparrow$	d	$\downarrow -\frac{1}{3}\hbar$	$-\frac{1}{3}e^{-}$

Total spin : 0
Total orbital : $-\hbar$
Total spin + orbital : $-\hbar$
Charge : e^{-}

Total j^{p}(spin + orbital) = $+1/2\hbar$, Neutral charge

$\rightarrow$

proton

spin		orbital	charge
$+\frac{1}{2}\hbar\uparrow$	u	$\uparrow +\frac{2}{3}\hbar$	$+\frac{2}{3}e^{+}$
$-\frac{1}{2}\hbar\downarrow$	u	$\uparrow +\frac{2}{3}\hbar$	$+\frac{2}{3}e^{+}$
$-\frac{1}{2}\hbar\downarrow$	d	$\downarrow -\frac{1}{3}\hbar$	$-\frac{1}{3}e^{-}$

Total spin : $-1/2\hbar$
Total orbital : $+\hbar$
Total spin + orbital : $+1/2\hbar$
Charge : e^{+}

$+$

π^{-}

spin		orbital	charge
$-\frac{1}{2}\hbar\downarrow$	$\bar{u}$	$\downarrow -\frac{2}{3}\hbar$	$-\frac{2}{3}e^{-}$
$+\frac{1}{2}\hbar\uparrow$	d	$\downarrow -\frac{1}{3}\hbar$	$-\frac{1}{3}e^{-}$

Total spin : 0
Total orbital : $-\hbar$
Total spin + orbital : $-\hbar$
Charge : e^{-}
π^{-} decays thru W^{-} boson into $e^{-}+\bar{\nu}_{e}$ or into $\mu^{-}+\bar{\nu}_{\mu}$

$$n_{excited} \rightarrow p + e^{-} + \bar{\nu}_{e} \quad \text{or} \quad n_{excited} \rightarrow p + \mu^{-} + \bar{\nu}_{\mu}$$

Note: The π^- meson has a zero spin and an orbital angular momentum of $(-\hbar\downarrow)$ in the excited neutron composition (see schema). While being emitted out from the third level thru the W^- boson, its orbital angular momentum $(-\hbar\downarrow)$ becomes the W^- boson spin. The $(-\hbar\downarrow)$ spin "splits" afterwards in a half between e^- and the anti neutrino $\overline{\nu}_e$ while W^- boson decays into them. While π^- meson is being emitted out, it leaves the total j^P of the excited neutron composition with a $(-\hbar\downarrow)$ short. To keep the "spin balance" of the original excited neutron, it is compensated by "adding" a $(-\hbar\downarrow)$ to the proton in the excited neutron composition that has a (Total spin + orbital : + 3/2 $\hbar$) before the π^- meson break away. A **spin flip** of one of the *up* quark from $(+\frac{1}{2}\hbar\uparrow)$ state to $(-\frac{1}{2}\hbar\downarrow)$ in this proton occurs spontaneously and it is equivalent to "adding" a $(-\hbar\downarrow)$ spin to the proton (see the excited neutron products in the last schema). The spin flip corresponds with the proton's content of (Total spin + orbital : + 1/2 $\hbar$) and also corresponds with the Pauli Exclusion Principle saying that no two identical particles with 1/2 $\hbar$ spin can occupy the same quantum state simultaneously.

b. The proton excited state: The proton's inner structure prior to its decaying thru the weak force during interaction with another subatomic particle is as following:

While exchanging pions within the nucleolus the proton absorbs a neutral meson π° that changes its inner binding energy configuration by a new generated π^+ meson (see the schema below) that reaches the third orbital level in a momentarily captured status which creates an **excited proton** with e^+ charge and 1/2$\hbar$ spin maintained.

proton

spin		orbital	charge
$+\frac{1}{2}\hbar\uparrow$	u	$\uparrow +\frac{2}{3}\hbar$	$+\frac{2}{3}e^+$
$-\frac{1}{2}\hbar\downarrow$	u	$\uparrow +\frac{2}{3}\hbar$	$+\frac{2}{3}e^+$
$-\frac{1}{2}\hbar\downarrow$	d	$\downarrow -\frac{1}{3}\hbar$	$-\frac{1}{3}e^-$

Total spin : $-1/2\,\hbar$
Total orbital : $+\hbar$
Total spin + orbital : $+1/2\,\hbar$
Charge : e^+

+

π°

spin		orbital	charge
$-\frac{1}{2}\hbar\downarrow$	d	$\downarrow -\frac{1}{3}\hbar$	$-\frac{1}{3}e^-$
$+\frac{1}{2}\hbar\uparrow$	$\overline{d}$	$\uparrow +\frac{1}{3}\hbar$	$+\frac{1}{3}e^+$

Total spin : 0
Total orbital : 0
Total spin + orbital : 0
Neutral charge :

→

excited proton

spin		orbital	charge
$+\frac{1}{2}\hbar\uparrow$	u	$\uparrow +\frac{2}{3}\hbar$	$+\frac{2}{3}e^+$
$-\frac{1}{2}\hbar\downarrow$	d	$\downarrow -\frac{1}{3}\hbar$	$-\frac{1}{3}e^-$
$-\frac{1}{2}\hbar\downarrow$	d	$\downarrow -\frac{1}{3}\hbar$	$-\frac{1}{3}e^-$

Total spin : $-1/2\,\hbar$
Total orbital : 0
Total spin + orbital : $-1/2\,\hbar$
Neutral charge :

π^+ positioned at the third level of the excited proton

spin		orbital	charge
$-\frac{1}{2}\hbar\downarrow$	u	$\uparrow +\frac{2}{3}\hbar$	$+\frac{2}{3}e^+$
$+\frac{1}{2}\hbar\uparrow$	$\overline{d}$	$\uparrow +\frac{1}{3}\hbar$	$+\frac{1}{3}e^+$

Total spin : 0
Total orbital : $+\hbar$
Total spin + orbital : $+\hbar$
Charge : e^+

Total j^P(spin + orbital) = $+1/2\,\hbar$
Charge : e^+

The **excited proton**

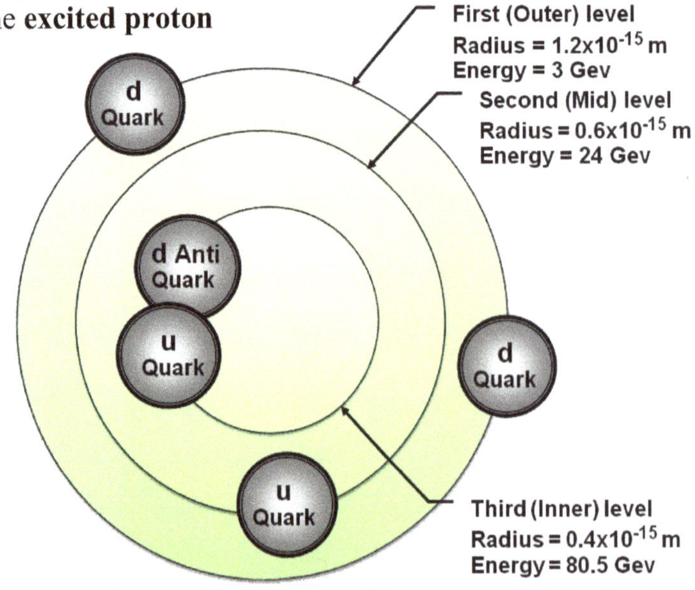

Excited Proton before the positive π meson at the third energy level is being emitted out and transformed into a Neutron.

excited proton

spin		orbital	charge
$+\frac{1}{2}\hbar\uparrow$	u	$\uparrow +\frac{2}{3}\hbar$	$+\frac{2}{3}e^+$
$-\frac{1}{2}\hbar\downarrow$	d	$\downarrow -\frac{1}{3}\hbar$	$-\frac{1}{3}e^-$
$-\frac{1}{2}\hbar\downarrow$	d	$\downarrow -\frac{1}{3}\hbar$	$-\frac{1}{3}e^-$

Total spin : $-1/2\,\hbar$
Total orbital : 0
Total spin + orbital : $-1/2\,\hbar$
Neutral charge :

π^+ positioned at the third level

spin		orbital	charge
$-\frac{1}{2}\hbar\downarrow$	u	$\uparrow +\frac{2}{3}\hbar$	$+\frac{2}{3}e^+$
$+\frac{1}{2}\hbar\uparrow$	$\bar{d}$	$\uparrow +\frac{1}{3}\hbar$	$+\frac{1}{3}e^+$

Total spin : 0
Total orbital : $+\hbar$
Total spin + orbital : $+\hbar$
Charge : e^+

Total j^P (spin + orbital) = $+1/2\,\hbar$
Charge : e^+

$\rightarrow$

neutron

spin		orbital	charge
$+\frac{1}{2}\hbar\uparrow$	u	$\uparrow +\frac{2}{3}\hbar$	$+\frac{2}{3}e^+$
$-\frac{1}{2}\hbar\downarrow$	d	$\downarrow -\frac{1}{3}\hbar$	$-\frac{1}{3}e^-$
$+\frac{1}{2}\hbar\uparrow$	d	$\downarrow -\frac{1}{3}\hbar$	$-\frac{1}{3}e^-$

Total spin : $+1/2\,\hbar$
Total orbital : 0
Total spin + orbital : $+1/2\,\hbar$
Neutral charge :

$+$

π^+

spin		orbital	charge
$-\frac{1}{2}\hbar\downarrow$	u	$\uparrow +\frac{2}{3}\hbar$	$+\frac{2}{3}e^+$
$+\frac{1}{2}\hbar\uparrow$	$\bar{d}$	$\uparrow +\frac{1}{3}\hbar$	$+\frac{1}{3}e^+$

Total spin : 0
Total orbital : $+\hbar$
Total spin + orbital : $+\hbar$
Charge : e^+
π^+ decays thru W^+ boson into $e^+ + \nu_e$
or into $\mu^+ + \nu_\mu$

$$p_{excited} \rightarrow n + e^+ + \nu_e \quad \text{or} \quad p_{excited} \rightarrow n + \mu^+ + \nu_\mu$$

Note: The π^+ meson has a zero spin and an orbital angular momentum of ($+\hbar\uparrow$) in the excited proton composition (see schema). While being emitted out from the third level thru the W^+ boson, its orbital angular momentum ($+\hbar\uparrow$) becomes the W^+ boson spin. The ($+\hbar\uparrow$) spin "splits" afterwards in a half between e^+ and the neutrino ν_e while W^+ boson decays into them. While π^+ meson is being emitted out, it leaves the total j^P of the excited proton composition with a ($+\hbar\uparrow$) short. To keep the "spin balance" of the original excited proton, it is compensated by "adding" a ($+\hbar\uparrow$) to the neutron that has a (Total spin + orbital : $-1/2\hbar\downarrow$) in the excited proton composition before π^+ meson break away. A **spin flip** of one of the *down* quark from ($-\frac{1}{2}\hbar\downarrow$) state to ($+\frac{1}{2}\hbar\uparrow$) in this neutron occurs spontaneously and it is equivalent to "adding" a ($+\hbar\uparrow$) to the neutron (see the excited proton products in the last schema). The spin flip corresponds with the neutron's content of (Total spin + orbital : $+1/2\hbar$) and also corresponds with the Pauli Exclusion Principle saying that no two identical particles with $1/2\,\hbar$ spin can occupy the same quantum state simultaneously.

As a generalization of this topic: There are other baryons besides the proton and the neutron in an excited state that can have a different meson besides the pi meson that reaches the third level of a relevant baryon in the course of inner binding energy configuration changes, and while these mesons are emitted out along with a γ photon as a released kinetic energy ~80[Gev] from the third level, they jointly form the a W^- boson or a W^+ boson.

The other mesons are:

K^- and K^+ mesons: ($K^- = \bar{u}s, K^+ = u\bar{s}$).

D_s^- and D_s^+ (strange D mesons): ($D_s^- = s\bar{c}, D_s^+ = \bar{s}c$).

B_c^- and B_c^+ (charmed B mesons): ($B_c^- = b\bar{c}, B_c^+ = \bar{b}c$).

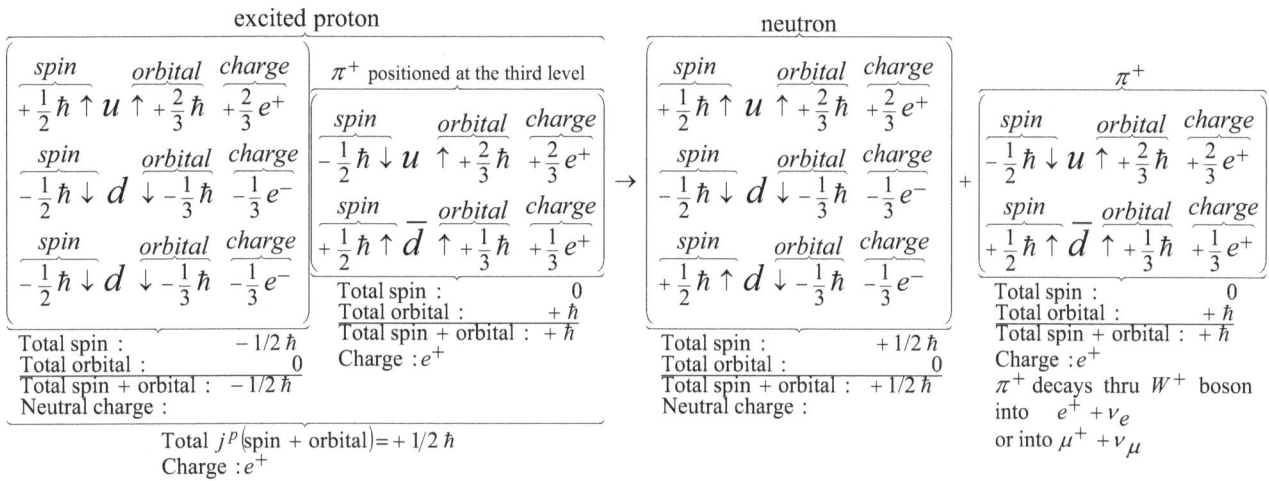

* Here are four experimental cases that corresponds with the subject discussed so far, using the excited proton $p_{excited}$ and the excited neutron $n_{excited}$.

a) The Cowan-Reines neutrino experiment (in this experiment the anti neutrino was discovered). It also called the **Inverse beta decay!**

$$\bar{\nu}_e + p \rightarrow n + e^+$$

An electron anti neutrino $\bar{\nu}_e$ collides with a proton: The outcome is a neutron and a positron e^+. What actually occurs is the following:

$$\bar{\nu}_e + p_{excited} \rightarrow n + e^+$$

$$\not{\bar{\nu}}_e + \underbrace{n + e^+ + \not{\nu}_e}_{p_{excited}} \rightarrow n + e^+$$

The anti neutrino $\bar{\nu}_e$ collides with an excited proton $p_{excited}$ and during the interaction the π^+ meson within the excited proton is being emitted out along with a γ photon that carries the released kinetic energy ~80[Gev] from the third energy level and jointly they form the W^+ boson that later decays into a positron e^+ and a neutrino ν_e. The anti neutrino $\bar{\nu}_e$ and the neutrino ν_e are mutually being annihilated and ultimately the outcome is a neutron accompanied by a positron:

b) Electron capture: A proton in a heavy nuclide formation absorbs an electron e^- from the lowest electron's orbital in the atom. The outcome is a neutron accompanied by a neutrino ν_e.

$$e^- + p \rightarrow n + \nu_e$$

What actually occurs is the following:

$$e^- + p_{excited} \rightarrow n + \nu_e$$

$$\not{e}^- + \underbrace{n + \not{e}^+ + \nu_e}_{p_{excited}} \rightarrow n + \nu_e$$

The electron e^- is being captured by an excited proton $p_{excited}$ from the first orbital in the atom, and during the interaction the π^+ meson within the excited proton is being emitted out from the third level along with a γ photon that carries the released kinetic energy ~80[Gev] from the third energy level and jointly they form the W^+ boson that later decays into a positron e^+ and a neutrino ν_e. The electron e^- and the positron e^+ are mutually being annihilated and ultimately the outcome is a neutron accompanied by a neutrino ν_e.

c) A muon anti neutrino $\bar{\nu}_\mu$ collides with a proton: The outcome is neutron and a muon μ^+.

$$\bar{\nu}_\mu + p \rightarrow n + \mu^+$$

What actually occurs is the following:

$$\bar{\nu}_\mu + p_{excited} \rightarrow n + \mu^+$$

$$\underbrace{\bar{\nu}_\mu + \underbrace{n + \mu^+ + \nu_\mu}_{p_{excited}}}_{} \rightarrow n + \mu^+$$

The muon anti neutrino $\bar{\nu}_\mu$ collides with an excited proton $p_{excited}$ and during the interaction the π^+ meson within the excited proton is being emitted out from the third level along with a γ photon that carries the released kinetic energy ~80[Gev] from the third energy level and jointly they form the W^+ boson that later decays into a muon μ^+ and a muon neutrino ν_μ. The muon anti neutrino $\bar{\nu}_\mu$ and the muon neutrino ν_μ are mutually being annihilated and the outcome is a neutron accompanied by a muon μ^+:

d) A muon neutrino ν_μ collides with a neutron: The outcome is a proton and a muon μ^-:

$$\nu_\mu + n \rightarrow p + \mu^-$$

What actually occurs is the following:

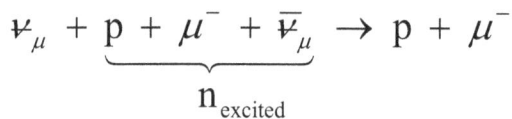

$$\nu_\mu + n_{excited} \rightarrow p + \mu^-$$

$$\nu_\mu + \underbrace{p + \mu^- + \bar{\nu}_\mu}_{n_{excited}} \rightarrow p + \mu^-$$

When muon neutrino ν_μ collides with an excited neutron the outcome is a proton and a muon μ^-:

The muon neutrino ν_μ collides with an excited neutron $n_{excited}$ and during the interaction the π^- meson within the excited neutron is being emitted out from the third level along with a γ photon that carries the released kinetic energy ~80[Gev] from the third energy level and jointly they form the W^- boson that later decays into a muon μ^- and an muon anti neutrino $\bar{\nu}_\mu$. The muon neutrino ν_μ and the muon antineutrino $\bar{\nu}_\mu$ are mutually being annihilated and ultimately the outcome is a proton accompanied by a muon μ^-:

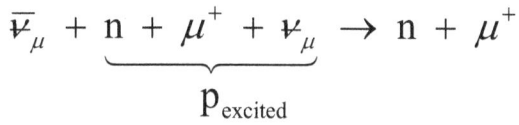

The Gluons:

The Gluons in this three orbital levels model represents the strong force particles exchanged between the different quarks within the proton or the neutron or other baryons or mesons that contains quarks. I am not using colors here to identify the types of the gluons. My intention is to show that their physical existence is a must according to this model. Please refer to page 178 for the color issue!

* **The first type gluon:** is being exchanged between the quark at the first orbital level and the quark at the second orbital level.

* **The second type gluon:** is being exchanged between the quark at the first orbital level and the quark at the third orbital level.

* **The third type gluon:** is being exchanged between the quark at the second orbital level and the quark at the third orbital level.

* **The forth type of gluon:** Logically the quarks at the first orbital level (the outer orbital that creates the surface of the proton or the neutron in this case) "are communicating" with the quarks in a similar position at the nearest nucleus particle so it requires a forth type of gluon that is being exchanged between them, and this creates the bounding of the particles in the nucleus (it refers to what is called "residual strong force").

* In addition there are four anti gluons types which make a total of eight gluon types!

The true nature of the neutrinos!

The γ photon emitted from the third energy level of a neutron or a proton along with charged Pi mesons thru the weak force process is actually a gluon that is fused into a pair of Ultra relativistic particles that travels nearly the speed of light c. By using the expression for the relativistic energy E of a particle with rest mass m and momentum p as given by: $E^2 = m^2c^4 + p^2c^2$, the energy of an Ultra relativistic particle comes mostly from its momentum $E = pc$ assuming $pc >> mc^2$.

This changes the characteristics of the pair of the Ultra relativistic particles which contains a large amount of energy E and merely negligible rest mass m and has a contracted radius due to the effects of relativistic laws while traveling nearly the speed of light c. An Ultra relativistic neutral particle with $1/2\hbar$ spin generated from a fused gluon, provides clues on the neutrino identity:

a. The neutrino ν is an Ultra relativistic state of a neutron traveling at nearly the speed of light c

b. The anti neutrino $\overline{\nu}$ is an Ultra relativistic state of an anti neutron travels nearly the light speed c

* The muon and tau neutrinos represents the differences between their traveling speeds near the limit of the speed of light in vacuum c !

Note: The kind of the neutron that decays to a proton thru the weak force interaction is an excited neutron presented previously on the "Weak Force decay thru $W^{\pm}$ boson mechanism" at page 147!

The **gluons** exchanged between the quarks within the baryons or being emitted out of the nucleus in the weak force process, are fused to produce neutrino/anti neutrino particles that consist the following (udd) and $(\overline{u}\overline{d}\overline{d})$ quarks in a form of a three Pi mesons $\pi^{\circ}, \pi^{+}, \pi^{-}$ or $\pi^{\circ}, \pi^{\circ}, \pi^{\circ}$ or in form of a second generation *quarks* (please see the new interpretation for the second generation quarks at chapter 15) as a **φ** meson that consists of a strange quark and a strange anti quark $s\overline{s}$: $(u\overline{u}d)$ and $(d\overline{d}\overline{d})$ or a **J/psi** meson that consists of a charm quark and a charm anti quark $c\overline{c}$: $(d\overline{d}u)$ and $(d\overline{d}\overline{u})$ that often participate in different kinds of interactions.

The Pi mesons generated from the gluons plays an important role in the following process:

* **The Pi mesons produced within the baryons:** are involved in the Strong force process!

* **The Pi mesons produced among the baryons** (from the fusion of the forth gluon type): are involved in the force that binds the proton and the neutron in the nucleus (Please see next page)

Note: The Pi mesons that are generated from fused gluons that produce neutrino/anti-neutrino are the source of creating new massive unstable particles during a collision of accelerated particles.

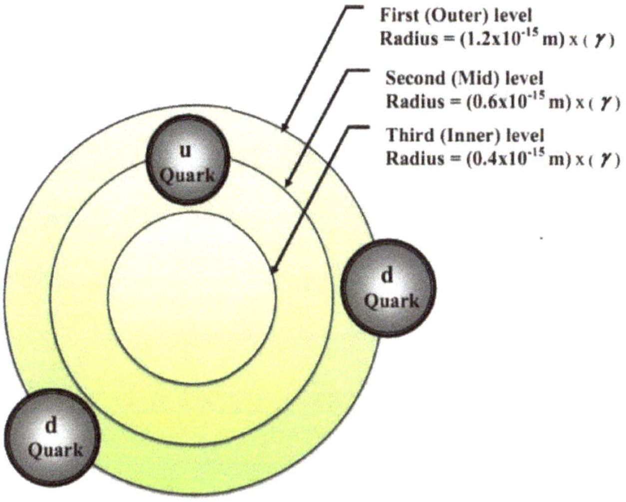

The neutrino (A relativistic state of a Neutron)

γ - The factor that creats the neutrino characteristics

$$\gamma = \sqrt{1 - v_N^2 / c^2}$$

v_N - Neutron traveling velocity near the speed of light.

c - speed of light in vacuum.

*The Energy levels depends on the γ factor. The neutrino is a state of an Ultra relativistic Neutron originated from a Gluon, and its Energy according to the relativistic rule comes mostly from its momentum $E = pc$, assuming $pc >> mc^2$. It contains a large amount of energy E and merely negligible rest mass m, and has a contracted radius!

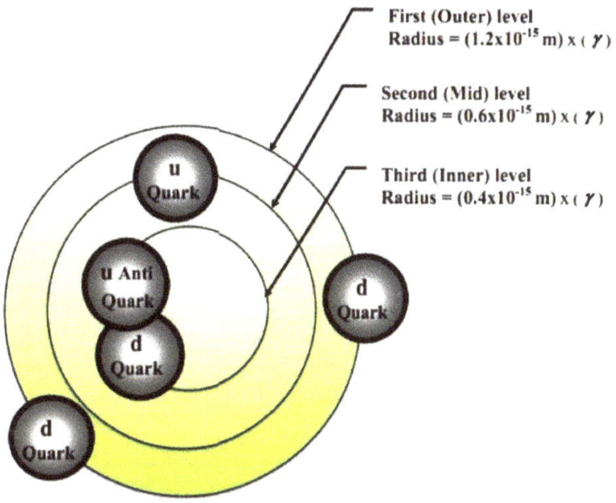

The Electron (A bound state of a neutrino and negative Pi meson)
Please see Chapter 16!

γ - The factor that creats the neutrino characteristics

$$\gamma = \sqrt{1 - v_N^2 / c^2}$$

v_N - Neutron traveling velocity near the speed of light.

c - speed of light in vacuum.

*The Energy levels depends on the γ factor. The neutrino is a state of an Ultra relativistic Neutron originated from a Gluon, and its Energy according to the relativistic rule comes mostly from its momentum $E = pc$, assuming $pc >> mc^2$. It contains a large amount of energy E and merely negligible rest mass m, and has a contracted radius!

* The force that binds the proton and the neutron in the nucleus is carried out by Pi Mesons in two forms:

a. The force is carried out by **exchanging** a neutral π° meson.

1. If π° meson is absorbed in the proton, the *antidown* quark of the π° meson annihilates the *down* quark of the proton, and the *down* quark of the π° meson occupies the vacant place in the proton.
2. If π° meson is absorbed in the neutron, the *antidown* quark of the π° meson annihilates one of the *down* quark of the neutron, and the *down* quark of the π° meson occupies the vacant place in the neutron.

b. The force is carried out by **exchanging** a neutral π° meson, but the neutral π° meson creates an **excited proton** or an **excited neutron** as it was explained previously.

1. If π° meson is absorbed in the proton and generates an excited state, the π^+ meson inside the excited proton is being emitted out (transforms the excited proton into a neutron) and it is being absorbed by another existing neutron (imagine as in the deuterium nucleus). Now the *antidown* quark of the π^+ meson annihilates one of the *down* quark of the existing neutron, and the *up* quark of the π^+ meson occupies a place at the second energy level in the formerly neutron, transforms the neutron into a proton.
2. If π° meson is absorbed in the neutron and generates an excited state, the π^- meson inside the excited neutron is being emitted out (transforms the excited neutron into a proton) and it is being absorbed by another existing proton (imagine as in the deuterium nucleus).Now the *antiup* quark of the π^- meson annihilates one of the *up* quark of the existing proton, and the *down* quark of the π^- meson occupies a place at the first energy level in the formerly proton, transforms the proton into a neutron.

14
The Higgs boson

The same technique that was used to describe the process that creates the top quark can be applied to what is supposed to be the Higgs Boson that its discovery announcement was made on 4th July 2012 at the CERN LHC particle Accelerator. The model of three energy levels in the proton presented in this theory predicts that when a proton and anti proton collides, there is a potential of finding a new particle at 128 GeV (or a bit less, it depends on the amount of energy that might be emitted as well which is not detected). This new particle generated from the proton/anti proton collision with about the mass of 128 Gev and a charge of a ($1/3\,e^{+}$) is suggesting that this particle is a new quark. The new quark decays simultaneously into a couple of particles. One of them is an *anti bottom* quark with a mass of ~ 4 Gev and charge of ($1/3\,e^{+}$) and the other will be a boson with a mass of ~ 125 Gev or a bit more (it is the difference between the new quark's mass and the *bottom* quark mass) and has a neutral charge. This boson is characterized as the Higgs boson and it decays in a couple of photons or a couple of quarks with opposite charges followed by particles jets and so forth.

Note: the experiment running at CERN is done by colliding proton versus proton, but the outcome should be the same for the collision that I am describing here on the proton side. Because there is symmetry between the outcomes, the collision between the proton and the anti proton generates similar particles with inverse electric charges and emitting directions. I am describing here the event of the proton decaying process in the collision which occurs simultaneously on the anti proton part.

The scenario: A proton and an anti proton collide in a particles accelerator involving an interaction with the second and the third energy levels within the proton.

At the third level, n=3: this is the lowest inner orbital in the proton. According to Eq. 252:

$$E_{n=3} = \frac{\vec{L}_{n=3} \times c \times (3\beta)}{2 \times \frac{r_p}{3}} \times \frac{1}{3}\frac{m_p}{m_e} = 1.292964123 \times 10^{-8}[J] = 80.7 \times 10^{9}[ev] = 80.7[Gev]$$

And for the **second orbital level n=2** in the proton. According to Eq. 254:

$$E_{n=2} = \frac{\vec{L}_{n=2} \times c \times (2\beta)}{2 \times \frac{r_p}{2}} \times \frac{1}{3}\frac{m_p}{m_e} = 3.831 \times 10^{-9}[J] = 2.3911 \times 10^{10}[ev] = 23.911[Gev]$$

So the total equivalent mass of the new particle generated in the collision on the proton part (a meson with a charge of e^{-} at the third level and two *up* quarks ($+2/3\,e^{+}$) at the second level) is as following:

$$E_{(\text{total energy})} = 1 \times 80.7[Gev] + 2 \times 23.911[Gev] = 128.5[Gev]$$

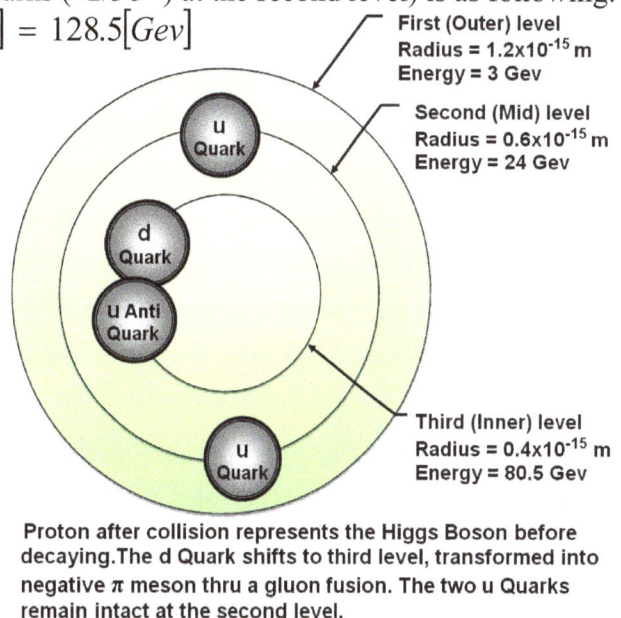

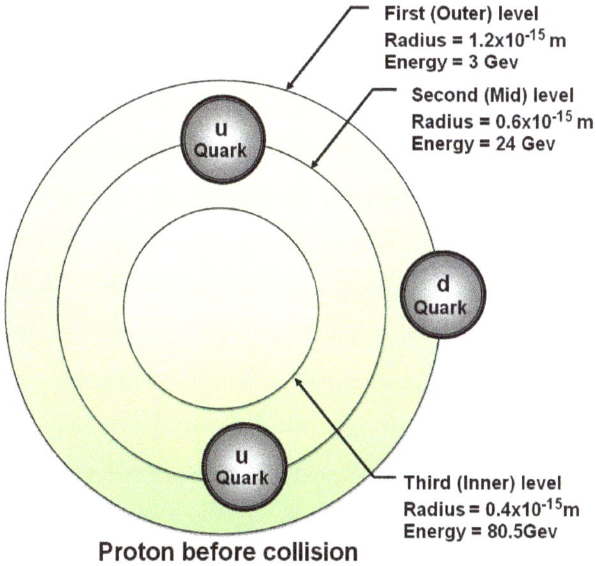

Proton before collision

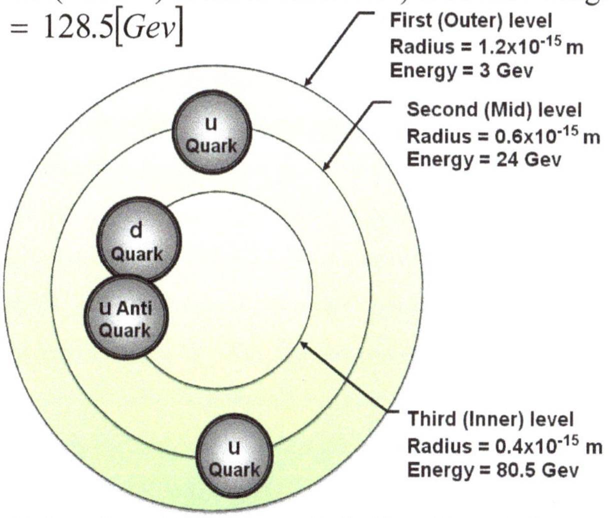

Proton after collision represents the Higgs Boson before decaying.The d Quark shifts to third level, transformed into negative π meson thru a gluon fusion. The two u Quarks remain intact at the second level.

Note: on the anti proton part there is a new particle generated with mass of ~128 Gev and charge of ($-1/3e^-$) that decays into a couple of particles. One of them is a *bottom* quark with a mass of ~ 4 Gev and charge of ($-1/3e^-$) the other is a boson with a mass of ~125 Gev and neutral!
The Higgs boson that decays into four leptons are described at page 186.
Here is a schema that shows the process:

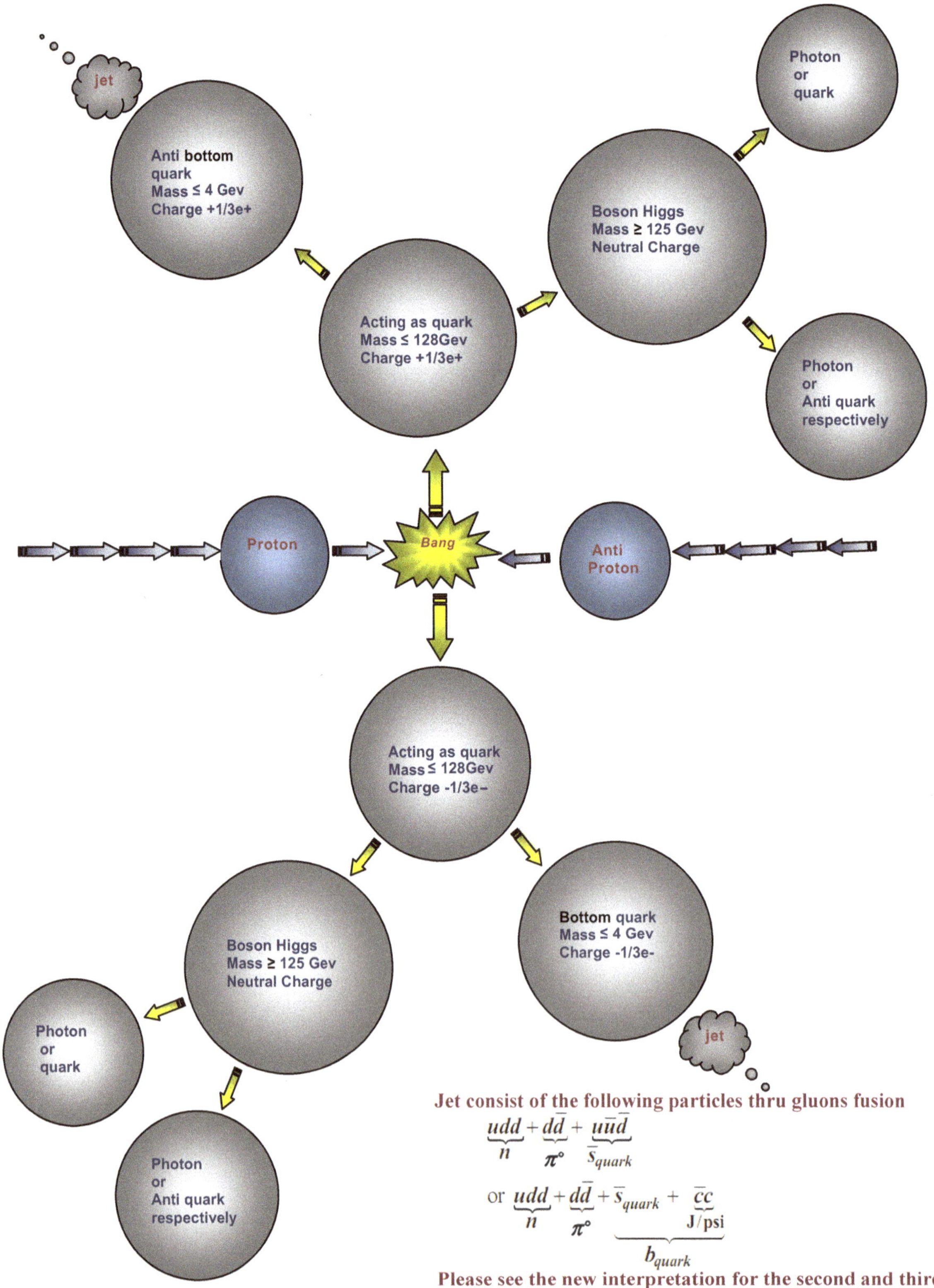

Jet consist of the following particles thru gluons fusion

$$\underbrace{udd}_{n}+\underbrace{d\bar{d}}_{\pi^\circ}+\underbrace{u\bar{u}d}_{\bar{s}_{quark}}$$

$$\text{or } \underbrace{udd}_{n}+\underbrace{d\bar{d}}_{\pi^\circ}+\underbrace{\bar{s}_{quark}+\underbrace{c\bar{c}}_{J/psi}}_{b_{quark}}$$

Please see the new interpretation for the second and third generation quarks next chapter 15, and Gluon nature page 150!

A schema shows the option to receive a boson that might be the Higgs Boson, in a proton anti proton collision.

Top quark / Anti Top quark, decays to Higgs boson (H) accompanied by a Charm quark c or anti Charm quark $\bar{c}$, then the Higgs Boson decays to b and $\bar{b}$ quarks.

***Please refer to chapter 15 'new interpretation for the second generation quarks' used here!**

1. A Quark within a baryon with a fractional electric charge which shifts during a collision event from the first energy level that permits a presence of particles with a third of the elementary charge or from the second energy level that permits a presence of particles with two thirds of the elementary electric charge, toward the third energy level that permits a presence of particles with a full unit of the elementary charge, will automatically coupled to a quark with a fractional electric charge from gluons hadronization that serves as a complementary part for achieving a full unit of the elementary electric charge as permitted at the third energy level.

2. In any event of baryons collision such as proton Vs proton or proton Vs anti proton, the process will absorb within a $s\bar{s}$ meson or a $c\bar{c}$ meson thru gluons hadronization, to enable their decaying into a different other particles. These absorbed particles are crucial for the decaying process, and sometimes it is needed to absorb more than one of them for different decaying options. This presentation shows how it works. Please see also the section on the gluon nature at page 152!

Note: The last process reminds the process of the OZI rule!

3. In order to be clear with the explanation as much as possible, I mark the quarks participating in the process according to their position at the three existing energy levels within the baryons, in order to track the replacement of the quarks' positions at the various energy levels, as following:

$d_{(1)}$ - The quark is positioned at the first energy level.

$u_{(2)}$ - The quark is positioned at the second energy level.

$d_{(3)}$ - The quark is positioned at the third energy level.

Proton before collision : $d_{(1)}u_{(2)}u_{(2)}$

$$\text{After the collision: } \underbrace{d_{(3)} + \overbrace{\bar{u}_{(3)}}^{\text{charge comp.}}}_{\pi^-}\ \underbrace{u_{(3)} + \overbrace{\bar{d}_{(3)}}^{\text{charge comp.}}}_{\pi^+}\ u_{(2)} + \overbrace{\underbrace{u_{(3)}\bar{u}_{(3)}d_{(3)}}_{s_{quark}}\ \underbrace{\bar{d}_{(3)}\bar{d}_{(3)}d_{(3)}}_{\bar{s}_{quark}}}^{\varphi}$$

* **Charge comp.** is an abbreviation for Charge complementary!
Please note that a $s\bar{s}$ meson is generated and absorbed within in the process during the impact!

$$\text{Prior decay : } \underbrace{d_{(3)}\bar{u}_{(3)}}_{\pi^-}\ \underbrace{u_{(3)}\bar{d}_{(3)}}_{\pi^+}\ u_{(2)} + \overbrace{\underbrace{\bar{u}_{(3)}d_{(3)}}_{\pi^-}\ \underbrace{\bar{d}_{(3)}u_{(3)}}_{\pi^+}\ d\bar{d}}^{\varphi} \rightarrow \underbrace{\pi^-\pi^+\pi^-\pi^+}\ + \underbrace{u_{(2)}\bar{d}_{(3)}d_{(3)}}_{c_{quark}}$$

Higgs boson pre decay status with 4 pions at the third level prior decaying into four leptons. Please see example at page 186!

Top quark decay option when four pions $\pi^-\pi^+\pi^-\pi^+$ are coupled to two pairs of $\nu_\mu\bar{\nu}_\mu$ neutrinos that are emitted out as four leptons from the third level revealing the Higgs boson presence :
The Top quark decay option process if the Higgs boson 'chooses' to decay to 4 leptons!

$$\underbrace{\underbrace{\pi^- + \nu_\mu}_{\mu^-} + \underbrace{\pi^+ + \bar{\nu}_\mu}_{\mu^+} + \underbrace{\pi^- + \nu_\mu}_{\mu^-} + \underbrace{\pi^+ + \bar{\nu}_\mu}_{\mu^+}} + c_{quark}$$

The revealed presence of the Higgs boson within the Top quark before it decays

Higgs prior decay to b and $\bar{b}$: $\underbrace{d_{(3)}\bar{u}_{(3)}}_{\pi^-}\ \underbrace{u_{(3)}\bar{d}_{(3)}}_{\pi^+}\ \underbrace{\bar{u}_{(3)}d_{(3)}}_{\pi^-}\ \underbrace{\bar{d}_{(3)}u_{(3)}}_{\pi^+} + \overbrace{\underbrace{u_{(3)}\bar{u}_{(3)}d_{(3)}}_{s_{quark}}\ \underbrace{\bar{d}_{(3)}d_{(3)}\bar{d}_{(3)}}_{\bar{s}_{quark}}}^{\varphi}$

The **Higgs boson decay option** to b and $\bar{b}$ (a $s\bar{s}$ meson is generated and absorbed within prior decaying process)!

$$\overbrace{\underbrace{d_{(3)}\bar{u}_{(3)}u_{(3)}}_{s_{quark}}\ \underbrace{\bar{d}_{(3)}\bar{u}_{(3)}u_{(3)}}_{\bar{s}_{quark}}\ d_{(3)}}^{b_{quark}}\ \overbrace{\bar{d}_{(3)}\ \underbrace{u_{(3)}\bar{u}_{(3)}d_{(3)}}_{s_{quark}}\ \underbrace{\bar{d}_{(3)}d_{(3)}\bar{d}_{(3)}}_{\bar{s}_{quark}}}^{\bar{b}_{quark}} \rightarrow \overbrace{s\bar{s}d}^{b_{quark}}\ \overbrace{s\bar{s}\bar{d}}^{\bar{b}_{quark}}$$

Anti Proton before collision : $\bar{d}_{(1)}\bar{u}_{(2)}\bar{u}_{(2)}$

After the collision : $\underbrace{\bar{d}_{(3)} + \overbrace{u_{(3)}}^{\text{charge comp.}}}_{\pi^+}\ \underbrace{\bar{u}_{(3)} + \overbrace{d_{(3)}}^{\text{charge comp.}}}_{\pi^-}\ \bar{u}_{(2)} + \overbrace{\underbrace{u_{(3)}\bar{u}_{(3)}d_{(3)}}_{s_{quark}}\ \underbrace{\bar{d}_{(3)}d_{(3)}\bar{d}_{(3)}}_{\bar{s}_{quark}}}^{\varphi}$

Please note that a $s\bar{s}$ meson is generated and absorbed within in the process during the impact!

Prior to decay: $\underbrace{\bar{d}_{(3)}u_{(3)}}_{\pi^+}\ \underbrace{\bar{u}_{(3)}d_{(3)}}_{\pi^-}\ \bar{u}_{(2)} + \overbrace{\underbrace{\bar{u}_{(3)}d_{(3)}}_{\pi^-}\ \underbrace{\bar{d}_{(3)}u_{(3)}}_{\pi^+}\ \bar{d}d}^{\varphi} \rightarrow \underbrace{\pi^-\pi^+\pi^-\pi^+}_{} + \underbrace{\bar{d}d\bar{u}_{(2)}}_{\bar{c}_{quark}}$

Higgs boson pre decay status with 4 pions at the third level prior decaying into four leptons. Please see example at page 186!

Top quark decay option when four pions $\pi^-\pi^+\pi^-\pi^+$ are coupled to two pairs of $\nu_\mu\bar{\nu}_\mu$ neutrinos that are emitted out as four leptons from the third level revealing the Higgs boson presence : **The Top quark decay option** process if the Higgs boson 'chooses' to decay to 4 leptons!

$$\underbrace{\underbrace{\pi^- + \nu_\mu}_{\mu^-} + \underbrace{\pi^+ + \bar{\nu}_\mu}_{\mu^+} + \underbrace{\pi^- + \nu_\mu}_{\mu^-} + \underbrace{\pi^+ + \bar{\nu}_\mu}_{\mu^+}}_{\text{The revealed presence of the Higgs boson within the Top quark before it decays}} + \bar{c}_{quark}$$

Higgs prior decay to b and $\bar{b}$: $\underbrace{u_{(3)}\bar{d}_{(3)}}_{\pi^+}\ \underbrace{d_{(3)}\bar{u}_{(3)}}_{\pi^-}\ \underbrace{\bar{u}_{(3)}d_{(3)}}_{\pi^-}\ \underbrace{\bar{d}_{(3)}u_{(3)}}_{\pi^+} + \overbrace{\underbrace{u_{(3)}\bar{u}_{(3)}d_{(3)}}_{s_{quark}}\ \underbrace{\bar{d}_{(3)}\bar{d}_{(3)}d_{(3)}}_{\bar{s}_{quark}}}^{\varphi}$

The **Higgs boson decay option** to b and $\bar{b}$ (a $s\bar{s}$ meson is generated and absorbed within prior decaying process)!

$$\overbrace{\underbrace{d_{(3)}\bar{u}_{(3)}u_{(3)}}_{s_{quark}}\ \underbrace{\bar{d}_{(3)}\bar{u}_{(3)}u_{(3)}}_{\bar{s}_{quark}}\ d_{(3)}}^{b_{quark}}\ \overbrace{\bar{d}_{(3)}\ \underbrace{u_{(3)}\bar{u}_{(3)}d_{(3)}}_{s_{quark}}\ \underbrace{\bar{d}_{(3)}\bar{d}_{(3)}d_{(3)}}_{\bar{s}_{quark}}}^{\bar{b}_{quark}} \rightarrow \overbrace{s\bar{s}d}^{b_{quark}}\ \overbrace{s\bar{s}\bar{d}}^{\bar{b}_{quark}}$$

Top quark / Anti Top quark, decays to Higgs boson (H) accompanied by a Charm quark c or anti Charm quark $\bar{c}$, then the Higgs Boson decays to b and $\bar{b}$ quarks.

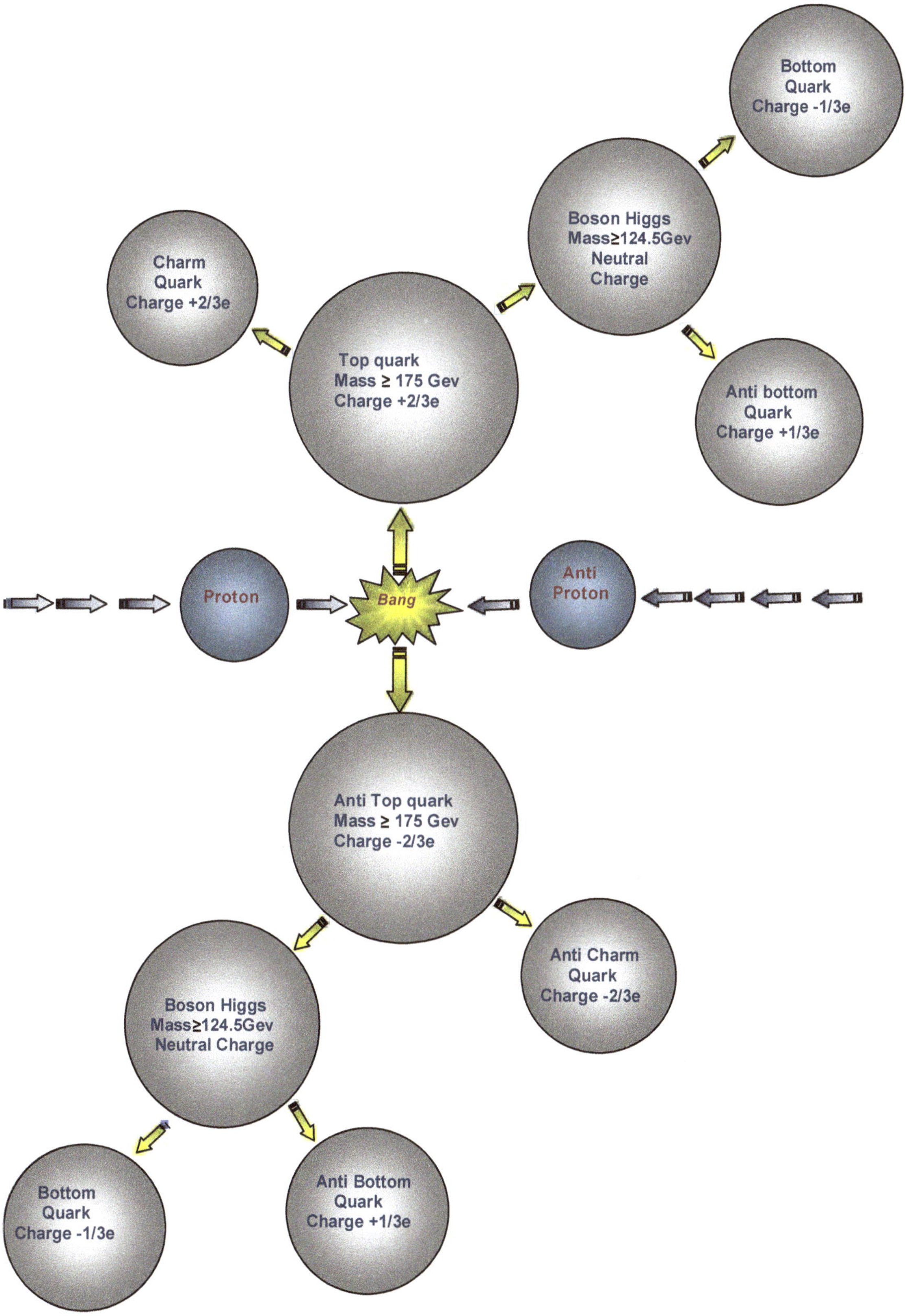

A prediction of a new neutral boson weighing 160Gev, probably a larger cousin of the Higgs boson that was discovered at CERN accelerator with the following properties:

Let's assume that all the three quarks in the proton that are tagged before the collision as: $\boldsymbol{d}_{(1)}\ \boldsymbol{u}_{(2)}\ \boldsymbol{u}_{(2)}$ are shifted after the collision to the third energy level and a new particle is formed weighing **240Gev** (this energy level equals **80Gev**), and a charge of e^+ (this is the total of their algebraic sum: {(-1/3e^-), (+2/3e^+), (+2/3e^+)}). This new particle will decay to a neutral boson that weighs **160Gev** accompanied by a W^+ **boson that weighs 80 Gev** (that makes the total of the original particle **240Gev),** and it is probably a larger cousin of the **Higgs boson 125Gev** that was discovered on **July 2012**.The neutral boson that weighs **160Gev** will decay into four leptons or a ***b*** **quark** and ***b*** **antiquark** and may be other option. **W^+ boson** will decay into a positron and an electron neutrino.

Note: Please keep in mind that I have described here one side of the collision from the aspect of a proton/proton collision type. If the collision is a proton/antiproton type, then on the antiproton side there will be: *anti* $\boldsymbol{d}_{(1)}$ *anti* $\boldsymbol{u}_{(2)}$ *anti* $\boldsymbol{u}_{(2)}$, that after collision it forms a new particle weighing **240 Gev**, and charge e^- (this is the total of their algebraic sum: {(+1/3e^+), (-2/3e^-), (-2/3e^-)}). This new particle will decay to a neutral boson that weighs **160Gev** accompanied by a negative **W boson** that weighs **80Gev.** The neutral boson that weighs **160Gev** will decay into four leptons or a ***b*** **quark** and ***b*** **antiquark** and may be other option. The negative **W boson** will decay into a muon and an anti muon neutrino.

At early 2007 analysis of collisions at **CDF** (The **Collider Detector at Fermilab's Tevatron**) seemed to support the existence of a Higgs like particle weighing **160Gev**!

The result of the analysis of collisions at **CDF** seems to be reliable.

15

A new Interpretation for the Second Generation Quarks

The quark model was initially proposed independently by physicists Murray Gell-Mann and George Zweig in 1964. Their model involved three flavors of quarks the up, down, and strange which carries properties such as spin and a fraction of elementary charge. There are six types of quarks, known as flavors: up, down (considered as first generation) , strange, charm (considered as second generation), and bottom, top (considered as third generation). The quarks and anti quarks that possessing opposite characteristics of electric charge and spin can be found only within hadrons such as baryons (proton, neutron…), and mesons (pi, k….). Quarks in the Standard Model of particle physics are involved in the fundamental interactions. Only the first-generation (up and down) quarks are usually occurred in nature. The quarks are currently presumed as being dimensionless point-like particles.

In a process of developing a sub atomic model I found that in reality there are less particles than those believed to be basic quarks since they possesses a fraction of the elementary charge, and are actually a complex particles consisting of sub atomic particles themselves. This means that experiments today observed the part of the elementary charge that these particles are revealing to us, while in fact they consist of three sub atomic particles where two of them neutralize each other due to their opposite electric charges thus preventing their detection. The quarks referred here are the *strange* quark and the *charm* quark, and based on specific assumptions, it is possible to formulate and achieve various decaying modes of the sub atomic particles theoretically, and especially the decay modes of the neutral K mesons that currently are explained by the **CP violation** analysis!

It seems that the *strange* quark is actually a particle that consists of three sub particles in two variations, two *down* quarks (-1/3e^-) and one *anti down* quark (1/3e^+) symbolized as $d\bar{d}d$ **or** one *up* quark (2/3e^+) one *antiup* quark (-2/3e^-) and one *down* quark (-1/3e^-) symbolized as $u\bar{u}d$. In the first option, one of the *down* quark is being neutralized by the *anti down* quark and therefore the *strange* quark appears to be having a charge of (-1/3e^-) which makes us believe that we deal with a basic quark besides the *up* quark and the *down* quark. The same in the second option, the *up* quark is being neutralized by the *antiup* quark and therefore the *strange* quark appears to be having a charge of (-1/3e^-).The *anti strange* quark consist of three sub particles, two *antidown* quarks (1/3e^+) and one *down* quark (-1/3e^-) symbolized as: $d\bar{d}\bar{d}$ or one *up* quark (2/3e^+) one *antiup* quark (-2/3e^-) and one *anti down* quark (1/3e^+) symbolized as: $u\bar{u}\bar{d}$ and again it appears to be having a charge of (1/3e^+). The mass of the *strange* quark is known to be heavier then the *down* quark and the *up* quark something around 0.54Gev (It is important to emphasize that part of its mass is used as a binding energy that holds the three inner particles together). This case holds for the *charm* quark as well (and apparently for the *bottom* quark as well). The *charm* quark is actually a quark that consist of three sub particles itself, one *up* quark (2/3e^+), one *strange* quark (-1/3e^-) and one *anti strange* quark (1/3e^+) symbolized as $c = s\bar{s}u$. One of the *strange* quark and the *anti strange* quark neutralizes each other due to their opposite electric charges and therefore are not detected! Hence, the *charm* quark appears to be having a charge of (2/3e^+). Additional option for the *charm* quark is $c = u\bar{u}u$ or $c = d\bar{d}u$. The *anti charm* quark consists of three particles, one *Anti up* quark (2/3e^-), one *strange* quark (-1/3e^-) and one *anti strange* quark (1/3e^+) symbolized as $\bar{c} = s\bar{s}\bar{u}$ or additional options $\bar{c} = u\bar{u}\bar{u}$ or $\bar{c} = d\bar{d}\bar{u}$, and again it appears to be having a charge of (-2/3e^-). Based on these assumptions it is possible to formulate and achieve theoretically a various decaying modes of the sub atomic particles into baryons or mesons. It will be done thru examples that show a systematic process. **Further decaying of baryons/mesons into leptons/photons is not introduced here except as a small reminder of the neutral Pi meson decaying into two photons!**

Table: Related Items and Symbols

d - *down quark*	$\bar{d}$ - *anti down quark*
u - *up quark*	$\bar{u}$ - *anti up quark*
s - *strange quark*	$s = d\bar{d}d$ **or** $s = u\bar{u}d$
$\bar{s}$ - *anti strange quark*	$\bar{s} = d\bar{d}\bar{d}$ **or** $\bar{s} = u\bar{u}\bar{d}$
c - *charm quark*	$c = s\bar{s}u$ **or** $c = u\bar{u}u$ **or** $c = d\bar{d}u$
$\bar{c}$ - *anti charm quark*	$\bar{c} = s\bar{s}\bar{u}$ **or** $\bar{c} = u\bar{u}\bar{u}$ **or** $\bar{c} = d\bar{d}\bar{u}$

Notes:

a. We substitute in the following equations: $s = d\bar{d}d$ *or* $s = u\bar{u}d$ *and* $\bar{s} = d\bar{d}\bar{d}$ *or* $\bar{s} = u\bar{u}\bar{d}$
$c = s\bar{s}u$ *and* $c = u\bar{u}u$ *or* $c = d\bar{d}u$ *and* $\bar{c} = u\bar{u}\bar{u}$ *or* $\bar{c} = d\bar{d}\bar{u}$!

b. The neutral K meson decay modes are currently explained by the **CP violation** methods. Following here I am presenting a different approach which matches the results obtained experimentally, but seemingly reveals what really happens in the process.

Presenting several decay modes using the assumptions

a. Decaying modes of baryon Λ° :

$$\Lambda^{\circ} = uds \rightarrow uddd\bar{d} \rightarrow udd + d\bar{d} \rightarrow n + \pi^{\circ}$$

$$\Lambda^{\circ} = uds \rightarrow udu\bar{u}d \rightarrow uud + \bar{u}d \rightarrow p + \pi^{-}$$

b. Decaying modes of baryons Ξ° and Ξ^{-}:

$$\Xi^{\circ} = uss \rightarrow usdd\bar{d} \rightarrow uds + d\bar{d} \rightarrow \Lambda^{\circ} + \pi^{\circ}$$

$$\Xi^{-} = dss \rightarrow dsu\bar{u}d \rightarrow uds + \bar{u}d \rightarrow \Lambda^{\circ} + \pi^{-}$$

c. Decaying modes of baryons Σ^{+} and Σ^{-}:

$$\Sigma^{+} = uus \rightarrow uudd\bar{d} \rightarrow uud + d\bar{d} \rightarrow p + \pi^{\circ}$$

$$\Sigma^{+} = uus \rightarrow uudd\bar{d} \rightarrow udd + u\bar{d} \rightarrow n + \pi^{+}$$

$$\Sigma^{-} = dds \rightarrow ddu\bar{u}d \rightarrow udd + \bar{u}d \rightarrow n + \pi^{-}$$

d. Decaying modes of baryon Ω^{-}:

$$\Omega^{-} = sss \rightarrow ssu\bar{u}d \rightarrow uds + \bar{u}s \rightarrow \Lambda^{\circ} + K^{-}$$

$$\Omega^{-} = sss \rightarrow ssu\bar{u}d \rightarrow uss + \bar{u}d \rightarrow \Xi^{\circ} + \pi^{-}$$

$$\Omega^{-} = sss \rightarrow ssdd\bar{d} \rightarrow dss + d\bar{d} \rightarrow \Xi^{-} + \pi^{\circ}$$

e. Decaying mode of meson K^{+} into **two pi** mesons.

$$K^{+} = u\bar{s} \rightarrow u\underbrace{u\bar{u}\bar{d}}_{\bar{s}} \rightarrow u\bar{d} + u\bar{u} \rightarrow \pi^{+} + \pi^{\circ}$$

f. Decaying modes of meson K^{+}, into **three pi** mesons.

This decay channel uses two sub phases as it will be shown depending on the amount of energy that is available at the interaction. First the *charm* quark followed by the *strange* quark.

The inverse process that generates K^+ is the following.

$$\pi^+ + \pi^\circ \rightarrow u\bar{d} + \bar{u}u \rightarrow u\underbrace{u\bar{u}\bar{d}}_{\bar{s}} \rightarrow u\bar{s} = K^+$$

The same **pi** mesons interaction undergoes fusion like that generated the K^+, but it uses a *different composition* of quarks per the assumption that decays as described in the schema that follows.

$$\pi^+ + \pi^\circ \rightarrow u\bar{d} + \bar{u}u \rightarrow \underbrace{uu\bar{u}}_{c}\bar{d} \rightarrow c\bar{d}$$

$$c\bar{d} \rightarrow s\bar{d} + \underbrace{u\bar{d}} \rightarrow u\bar{d} + \bar{u}d + u\bar{d} = \pi^+ + \pi^- + \pi^+$$

decaying modes thru W boson in schema

$$\left(c \rightarrow s + u\bar{d} \;\; and \;\; s \rightarrow u + \bar{u}d \right)$$

$$c\bar{d} \rightarrow \underbrace{s\bar{d}}_{\bar{K}^\circ} + \underbrace{u\bar{d}}_{\pi^+} \rightarrow \underbrace{u\bar{u}d}_{s}\bar{d}u\bar{d} \rightarrow u\bar{d} + \bar{u}d + u\bar{d} = \pi^+ + \pi^- + \pi^+$$

Or in alternative decaying configuration mode

$$c\bar{d} \rightarrow \underbrace{s\bar{d}}_{\bar{K}^\circ} + \underbrace{u\bar{d}}_{\pi^+} \rightarrow \underbrace{u\bar{u}d}_{s}\bar{d}u\bar{d} \rightarrow u\bar{u} + d\bar{d} + u\bar{d} = \pi^\circ + \pi^\circ + \pi^+$$

g. Decaying mode of meson K^- into **two pi** mesons.

$$K^- = \bar{u}s \rightarrow \bar{u}\underbrace{u\bar{u}d}_{s} \rightarrow \bar{u}d + \bar{u}u \rightarrow \pi^- + \pi^\circ$$

h. Decaying modes of meson K^-, into **three pi** mesons.

This decay channel uses two sub phases in a row.

The inverse process that generates K^- is the following

$$\pi^- + \pi^\circ \rightarrow \bar{u}d + \bar{u}u \rightarrow \bar{u}\underbrace{\bar{u}ud}_{s} \rightarrow \bar{u}s = K^-$$

The same **pi** mesons interaction undergoes fusion like that generated the K^-, uses a *different composition* of quarks, and decays as described in the schema that follows.

$$\pi^- + \pi^\circ \rightarrow \bar{u}d + \bar{u}u \rightarrow \underbrace{\bar{u}\bar{u}u}_{\bar{c}}d \rightarrow \bar{c}d$$

$$\bar{c}d \rightarrow \bar{s}d + \underbrace{\bar{u}d} \rightarrow \bar{u}d + u\bar{d} + \bar{u}d = \pi^- + \pi^+ + \pi^-$$

decaying modes thru W boson in schema

$$\left(\bar{c} \rightarrow \bar{s} + \bar{u}d \;\; and \;\; \bar{s} \rightarrow \bar{u} + u\bar{d} \right)$$

$$\bar{c}d \rightarrow \underbrace{\bar{s}d}_{K^\circ} + \underbrace{\bar{u}d}_{\pi^-} \rightarrow \underbrace{\bar{u}u\bar{d}}_{\bar{s}}d\bar{u}d \rightarrow \bar{u}d + u\bar{d} + \bar{u}d = \pi^- + \pi^+ + \pi^-$$

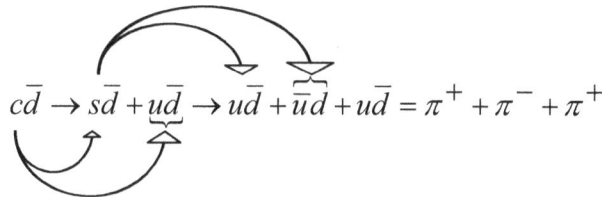

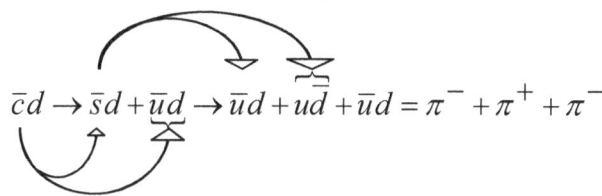

i. Decaying modes of meson K° into **two pi** mesons.

$$K^\circ = d\bar{s} \rightarrow d\bar{u}u\bar{d} \rightarrow \bar{u}d + u\bar{d} \rightarrow \pi^- + \pi^+$$

$$K^\circ = d\bar{s} \rightarrow d\bar{u}u\bar{d} \rightarrow u\bar{u} + d\bar{d} \rightarrow \pi^\circ + \pi^\circ$$

j. Decaying mode of meson $\overline{K}^\circ$ into **two pi** mesons.

$$\overline{K}^\circ = \bar{d}s \rightarrow \bar{d}u\bar{u}d \rightarrow u\bar{d} + \bar{u}d \rightarrow \pi^+ + \pi^-$$

k. Decaying modes of meson $\overline{K}^\circ$, into **three pi** mesons.

This decay channel uses two sub phases in a row.

The inverse process that generates $\overline{K}^\circ$ is the following

$$\pi^+ + \pi^- \rightarrow u\bar{d} + \bar{u}d \rightarrow \bar{d}\underbrace{u\bar{u}d}_{s} \rightarrow \bar{d}s = \overline{K}^\circ$$

The same **pi** mesons interaction that generated the $\overline{K}^\circ$, uses a *different composition* of quarks and decays as described in the schema that follows.

$$\pi^+ + \pi^- \rightarrow u\bar{d} + \bar{u}d \rightarrow u\underbrace{\bar{u}d\bar{d}}_{\bar{c}} \rightarrow u\bar{c}\,.$$

$$u\bar{c} \rightarrow u\bar{s} + \bar{u}d \rightarrow u\bar{u} + u\bar{d} + \bar{u}d = \pi^\circ + \pi^+ + \pi^-$$

decaying modes thru W boson in schema

$$\left(\bar{c} \rightarrow \bar{s} + \bar{u}d \;\; and \;\; \bar{s} \rightarrow \bar{u} + u\bar{d}\right)$$

$$u\bar{c} \rightarrow \underbrace{u\bar{s}}_{K^+} + \underbrace{\bar{u}d}_{\pi^-} \rightarrow u\underbrace{\bar{u}u\bar{d}}_{\bar{s}}\bar{u}d \rightarrow u\bar{u} + u\bar{d} + \bar{u}d = \pi^\circ + \pi^+ + \pi^-$$

Or in alternative decaying configuration mode

$$u\bar{c} \rightarrow \underbrace{u\bar{s}}_{K^+} + \underbrace{\bar{u}d}_{\pi^-} \rightarrow u\underbrace{\bar{u}u\bar{d}}_{\bar{s}}\bar{u}d \rightarrow u\bar{u} + u\bar{u} + d\bar{d} = \pi^\circ + \pi^\circ + \pi^\circ$$

l. Antiproton/proton collision generates:- $2\pi^- + 2\pi^+ + \pi^\circ$.

$$\bar{p} + p \rightarrow \bar{u}\bar{u}\bar{d} + uud \rightarrow \underbrace{\bar{u}\bar{d}d}_{\bar{c}} + \underbrace{\bar{u}uu}_{c} \rightarrow \bar{c}c \rightarrow \underline{su\bar{d}} + \underline{\bar{s}\bar{u}d} \rightarrow s\bar{s} + \underbrace{u\bar{d}}_{\pi^+} + \underbrace{\bar{u}d}_{\pi^-}$$

$$\rightarrow \underbrace{u\bar{u}d}_{s} + \underbrace{\bar{u}u\bar{d}}_{\bar{s}} + u\bar{d} + \bar{u}d \rightarrow \underbrace{u\bar{u}}_{\pi^\circ} + \underbrace{\bar{u}d}_{\pi^-} + \underbrace{u\bar{d}}_{\pi^+} + \underbrace{u\bar{d}}_{\pi^+} + \underbrace{\bar{u}d}_{\pi^-} \rightarrow 2\pi^- + 2\pi^+ + \pi^\circ$$

m. Meson K^- and proton collision generates:- $\Omega^- + K^+ + \overline{K}^\circ$.(from the Ω^- discovery in 1964)

Note: The K^- meson here decays into **three pi** mesons:- $K^- \rightarrow \bar{u}d + u\bar{d} + \bar{u}d$ (see paragraph ***h***)

$$\underbrace{\bar{u}du\bar{d}\bar{u}d}_{K^-}\underbrace{uud}_{p} \rightarrow \underbrace{d\bar{d}u}_{c}\underbrace{u\bar{u}u}_{c}\bar{u}dd \rightarrow \underbrace{s\bar{s}u}_{c}\underbrace{s\bar{s}u}_{c}\bar{u}dd \rightarrow ss\bar{s}\bar{s}\underbrace{u\bar{u}d}_{s}ud \rightarrow \underbrace{sss}_{\Omega^-} + \underbrace{u\bar{s}}_{K^+} + \underbrace{d\bar{s}}_{K^\circ}$$

n. Discovering of Ω^- at Brookhaven National Laboratory in1964:- $\Xi^\circ + \gamma\gamma + \Lambda^\circ + \pi^- + p$.

* Please follow the arrows at the **omega minus decay sequence** in the schema below, and see the explanation of how the neutral pion decays to four leptons through the couple of photons at page 186!

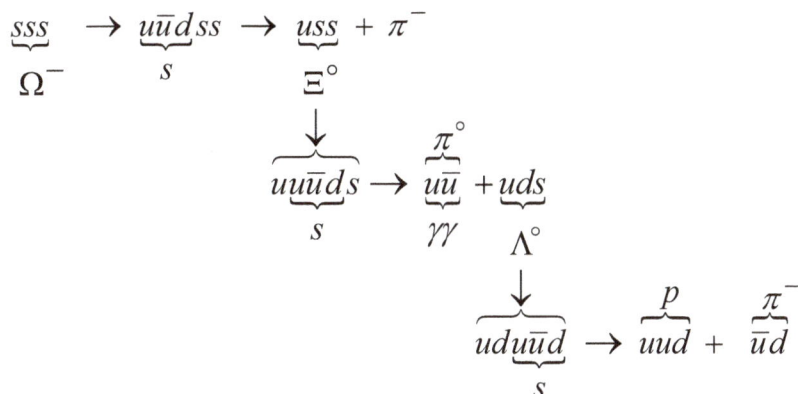

$$\underbrace{sss}_{\Omega^-} \rightarrow \underbrace{u\bar{u}d}_{s}ss \rightarrow \underbrace{uss}_{\Xi^\circ} + \pi^-$$

$$\downarrow$$

$$\overbrace{u\underbrace{u\bar{u}d}_{s}s} \rightarrow \overbrace{\underbrace{u\bar{u}}_{\gamma\gamma}}^{\pi^\circ} + \underbrace{uds}_{\Lambda^\circ}$$

$$\downarrow$$

$$\overbrace{ud\underbrace{u\bar{u}d}_{s}} \rightarrow \overbrace{uud}^{p} + \overbrace{\bar{u}d}^{\pi^-}$$

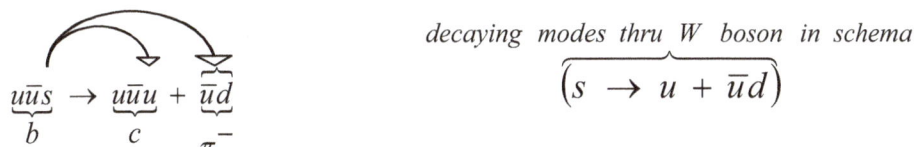

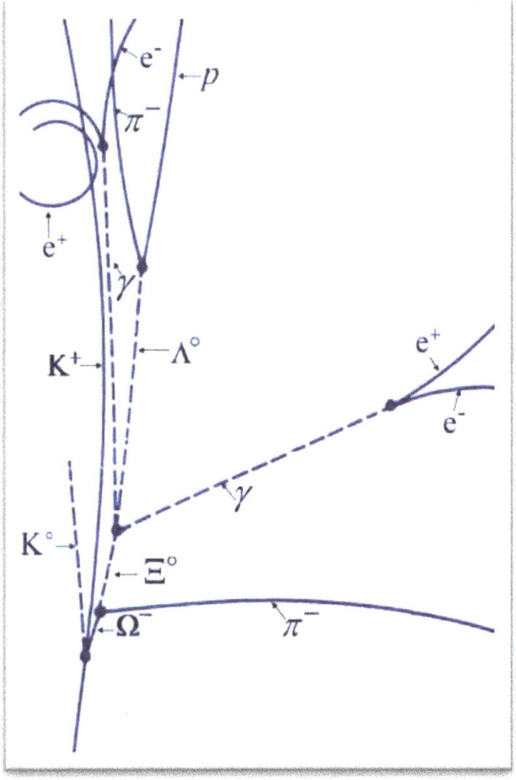

* The bottom quark ***b*** (is considered as a third generation quark) has the following combinations: $b = u\bar{u}s$ or $b = s\bar{s}s$ or $b = c\bar{c}s$, and the anti bottom quark is $\bar{b} = u\bar{u}\bar{s}$ or $\bar{b} = s\bar{s}\bar{s}$ or $\bar{b} = c\bar{c}\bar{s}$.

Note: Please notice that the *bottom* quark has a charge of [-1/3 *e*] manifested by its strange quark.
o. Decaying modes of quark $b\,[bottom]$:

$$\underbrace{u\bar{u}s}_{b} \rightarrow \underbrace{u\bar{u}u}_{c} + \underbrace{\bar{u}d}_{\pi^-} \qquad \overbrace{(s \rightarrow u + \bar{u}d)}^{decaying\ modes\ thru\ W\ boson\ in\ schema}$$

The bottom quark ***b*** can also decay into the following mode:

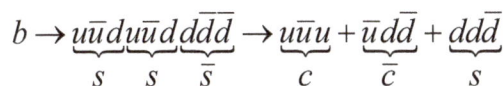

$$b \rightarrow \underbrace{u\bar{u}d}_{s}\underbrace{u\bar{u}d}_{s}\underbrace{d\bar{d}\bar{d}}_{\bar{s}} \rightarrow \underbrace{u\bar{u}u}_{c} + \underbrace{\bar{u}d\bar{d}}_{\bar{c}} + \underbrace{dd\bar{d}}_{s}$$

The new interpetation defines the strange quark as being actually a baryon consisting of three sub particles itself. for instance the *strange* quark consists of two *down* quarks (-1/3 e^-) and one *antidown* quark (+1/3 e^+) symbolized $s = dd\bar{d}$, and the *anti strange* quark consists of two *anti down* quarks (+1/3 e^+) and one *down* quark (-1/3 e^-) symbolized $\bar{s} = \bar{d}\bar{d}d$.
The early development in physics that led to the concept of the quarks model was the *Eight fold* way model which introduces a sub atomic particles classification arranged symetricly in a layout of super multiplets (octets, decuplets) based on the S(U)3 theory.
This new interpretation presenting this sub atomic clasiffication in a way that apperntly provides the same results and more!
For the calculation purpose I will use the the following known terms and equations:
I_3 : - a term symbolized the Isospin concept of the sub atomic particles.
$I_3 = 1/2\left[(nu - n\bar{u}) - (nd - n\bar{d})\right]$
nu : - the quantity number $[n]$ of the *up* $[u]$quarks in the relevant baryon or meson.
$n\bar{u}$: - the quantity number $[n]$ of the *anti up* $[\bar{u}]$quarks in the relevant baryon or meson.
nd : - the quantity number $[n]$ of the *down* $[d]$quarks in the relevant baryon or meson.
$n\bar{d}$: - the quantity number $[n]$ of the *anti down* $[\bar{d}]$quarks in the relevant baryon or meson.
Q: - the electric charge of a baryons or a meson
$Q = I_3 + 1/2(A + S)$
A: - baryon number.
S: - strangeness (defines how many strange quarks are in a baryon or a meson, gets a positive or negative values depending on the strange quark charge).
First lets us present the Decuplet of the heavy baryons, by using the terms and equations introduced:

$\Delta^- = ddd;\ \ A = 1;\ \ S = 0; \rightarrow I_3 = 1/2[(0-0)-(3-0)] = -3/2;\ \ Q = -3/2 + 1/2(1+0) = -1$

$\Delta^\circ = udd;\ \ A = 1;\ \ S = 0; \rightarrow I_3 = 1/2[(1-0)-(2-0)] = -1/2;\ \ Q = -1/2 + 1/2(1+0) = 0$

$\Delta^+ = uud;\ \ A = 1;\ \ S = 0; \rightarrow I_3 = 1/2[(2-0)-(1-0)] = 1/2;\ \ Q = 1/2 + 1/2(1+0) = 1$

$\Delta^{++} = uuu;\ \ A = 1;\ \ S = 0; \rightarrow I_3 = 1/2[(3-0)-(0-0)] = 3/2;\ \ Q = 3/2 + 1/2(1+0) = 2$

$\Sigma^{*-} = dds = dddd\bar{d};\ \ A = 2;\ \ S = -1; \rightarrow I_3 = 1/2[(0-0)-(4-1)] = -3/2;\ \ Q = -3/2 + 1/2(2-1) = -1$

Note: Here is the difference between the previous method and the new one according the new interpretation of the *strange* quark as a baryon.
For all the particles that consist of *strange* quarks, the A which represents the quantity of the baryons, will be considered as the sum of the original baryon plus the *strange* quark within as by the new interpretation for being a baryon itself, for instance:

$\Sigma^{*-} = dds$ is considered as one baryon, and a $s = dd\bar{d}$ within is another, which makes a total of A=2!

$\Xi^{*-} = dss$ is considered as one baryon, and two $s = dd\bar{d}$ are the others, which makes a total of A=3!

$\Sigma^{*\circ} = uds = uddd\bar{d};\ \ A = 2;\ \ S = -1; \rightarrow I_3 = 1/2[(1-0)-(3-1)] = -1/2;\ \ Q = -1/2 + 1/2(2-1) = 0$

$\Sigma^{*+} = uus = uudd\bar{d};\ \ A = 2;\ \ S = -1; \rightarrow I_3 = 1/2[(2-0)-(2-1)] = 1/2;\ \ Q = 1/2 + 1/2(2-1) = 1$

$\Xi^{*-} = dss = ddd\bar{d}dd\bar{d};\ \ A = 3;\ \ S = -2; \rightarrow I_3 = 1/2[(0-0)-(5-2)] = -3/2;\ \ Q = -3/2 + 1/2(3-1) = -1$

$\Xi^{*\circ} = uss = udd\bar{d}dd\bar{d};\ \ A = 3;\ \ S = -2; \rightarrow I_3 = 1/2[(1-0)-(4-2)] = -1/2;\ \ Q = -1/2 + 1/2(3-2) = 0$

$\Omega^{*-} = sss = dd\bar{d}dd\bar{d}dd\bar{d};\ \ A = 4;\ \ S = -3; \rightarrow I_3 = 1/2[(0-0)-(6-3)] = -3/2; Q = -3/2 + 1/2(4-3) = -1$

The Decuplet of the heavy baryons schema is presented at the next page.

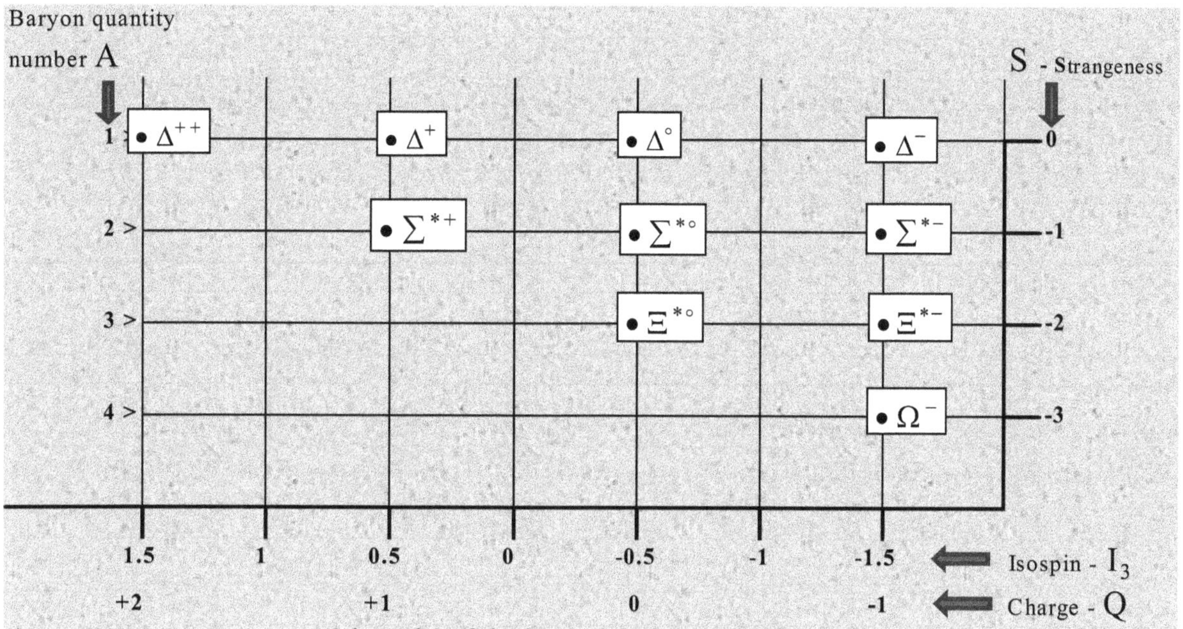

decuplet of the heavy baryons (spin 3/2)

Now let us present the octet of the semi stable baryons, by using the terms and equations introduced:

$p_{proton} = uud;\ A = 1;\ S = 0; \rightarrow I_3 = 1/2[(2-0)-(1-0)] = 1/2;\ Q = 1/2 + 1/2(1-0) = 1$

$n_{neutron} = udd;\ A = 1;\ S = 0; \rightarrow I_3 = 1/2[(1-0)-(2-0)] = -1/2;\ Q = -1/2 + 1/2(1-0) = 0$

$\Sigma^{-}{}_{Baryon} = dds = dddd\bar{d};\ A = 2;\ S = -1; \rightarrow I_3 = 1/2[(0-0)-(4-1)] = -3/2;\ Q = -3/2 + 1/2(2-1) = -1$

$\Sigma^{\circ}{}_{Baryon} = uds = uddd\bar{d};\ A = 2;\ S = -1; \rightarrow I_3 = 1/2[(1-0)-(3-1)] = -1/2;\ Q = -1/2 + 1/2(2-1) = 0$

$\Sigma^{+}{}_{Baryon} = uus = uudd\bar{d};\ A = 2;\ S = -1; \rightarrow I_3 = 1/2[(2-0)-(2-1)] = 1/2;\ Q = 1/2 + 1/2(2-1) = 1$

$\Lambda^{\circ}{}_{Baryon} = uds = uddd\bar{d};\ A = 2;\ S = -1; \rightarrow I_3 = 1/2[(1-0)-(3-1)] = -1/2;\ Q = -1/2 + 1/2(2-1) = 0$

$\Xi^{\circ}{}_{Baryon} = uss = udd\bar{d}dd\bar{d};\ A = 3;\ S = -2; \rightarrow I_3 = 1/2[(1-0)-(4-2)] = -1/2;\ Q = -1/2 + 1/2(3-2) = 0$

$\Xi^{-}{}_{Baryon} = dss = ddd\bar{d}dd\bar{d};\ A = 3;\ S = -2; \rightarrow I_3 = 1/2[(0-0)-(5-2)] = -3/2;\ Q = -3/2 + 1/2(3-2) = -1$

The octet of the semi stable baryons classification scheme is presented here below:

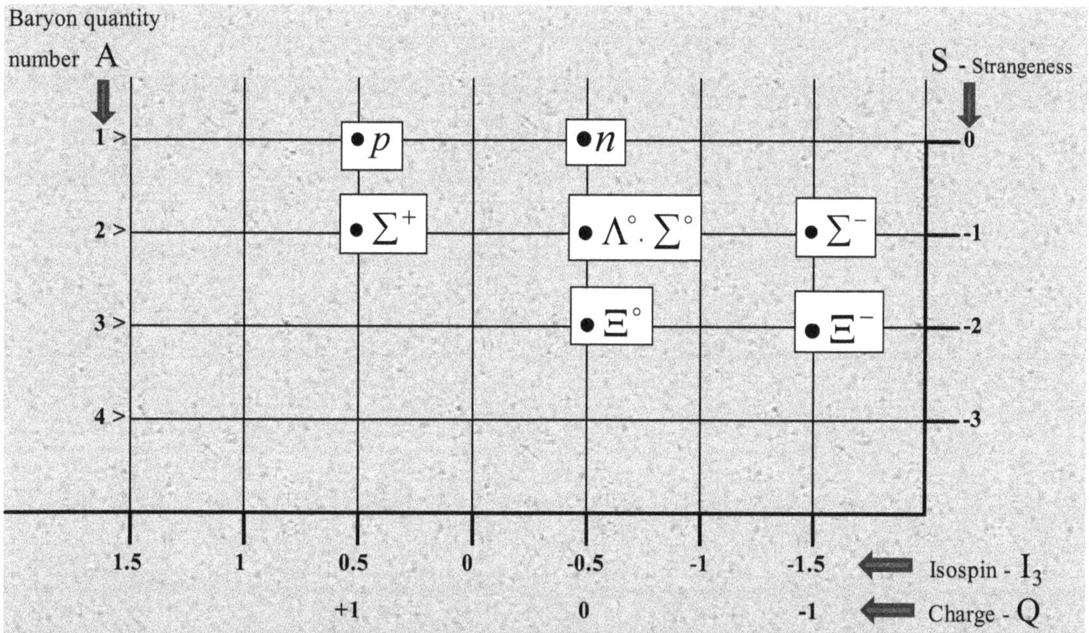

octet of the semi stable baryons (spin 1/2)

And as a last example let us present the octet of the semi stable mesons, by using the terms and equations introduced:

$K^{\circ}{}_{Kaon} = d\bar{s} = d\overline{dd}d;\ A=-1;\ S=1; \rightarrow I_3 = 1/2\left[(0-0)-(2-2)\right]=0;\ Q=0+1/2(-1+1)=0$

$K^{+}{}_{Kaon} = u\bar{s} = u\overline{dd}d;\ A=-1;\ S=1; \rightarrow I_3 = 1/2\left[(1-0)-(1-2)\right]=1;\ Q=1+1/2(-1+1)=1$

$K^{-}{}_{Kaon} = \bar{u}s = \bar{u}dd\bar{d};\ A=1;\ S=-1; \rightarrow I_3 = 1/2\left[(0-1)-(2-1)\right]=-1;\ Q=-1+1/2(1-1)=-1$

$\overline{K}^{\circ}{}_{Kaon} = \bar{d}s = \bar{d}dd\bar{d};\ A=1;\ S=-1; \rightarrow I_3 = 1/2\left[(0-0)-(2-2)\right]=0;\ Q=0+1/2(1-1)=0$

$\pi^{-}{}_{pion} = \bar{u}d;\ A=0;\ S=0; \rightarrow I_3 = 1/2\left[(0-1)-(1-0)\right]=-1;\ Q=-1+1/2(0+0)=-1$

$\pi^{\circ}{}_{pion} = d\bar{d};\ A=0;\ S=0; \rightarrow I_3 = 1/2\left[(0-0)-(1-1)\right]=0;\ Q=0+1/2(0+0)=0$

$\pi^{+}{}_{pion} = u\bar{d};\ A=0;\ S=0; \rightarrow I_3 = 1/2\left[(1-0)-(0-1)\right]=1;\ Q=1+1/2(0+0)=1$

$\eta^{\circ}{}_{meson} = d\bar{d};\ A=0;\ S=0; \rightarrow I_3 = 1/2\left[(0-0)-(1-1)\right]=0;\ Q=0+1/2(0+0)=0$

The octet of the semi stable mesons classification scheme is presented here below:

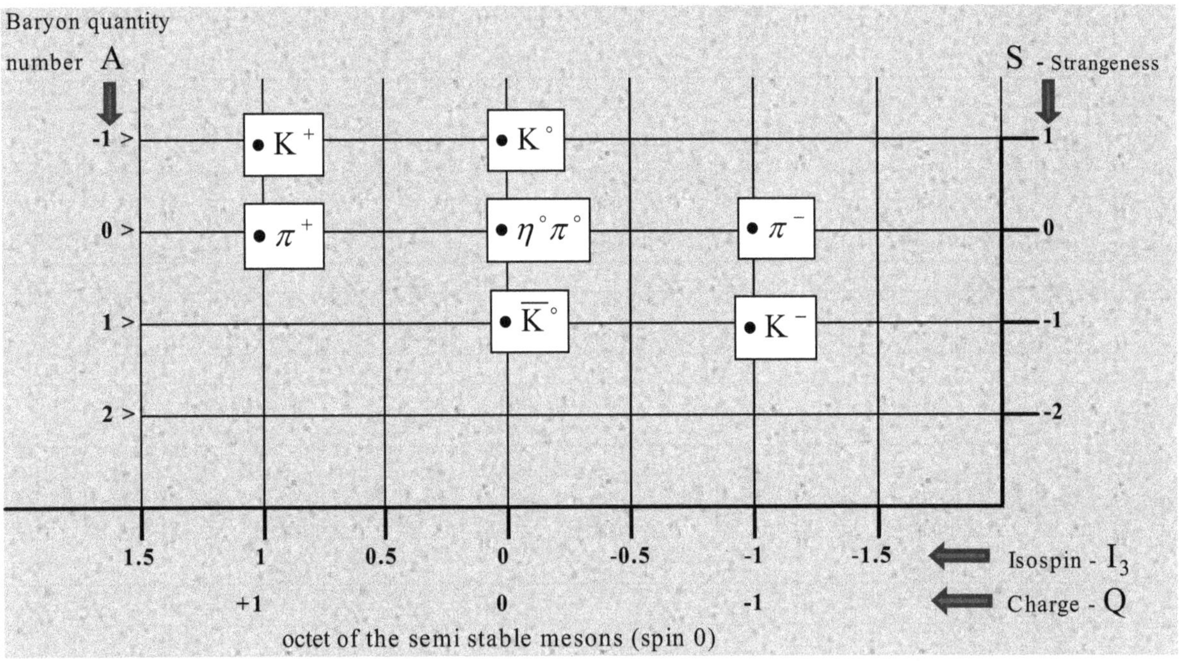

Another insight revealed from the new interpretation of the strange quark came from analyzing the **Gell-Mann–Okubo mass formula** given here (please see the schema below):

$$m_{\Delta^+} - m_{\Sigma^{*+}} = m_{\Sigma^{*\circ}} - m_{\Xi^{*\circ}} = m_{\Xi^{*-}} - m_{\Omega^-}$$

These equalities represent the quantitative differences between the average masses of the particles that occupy the relevant multiplets defined by their Strangeness versus each Charge and Isospin I_3 columns. The chart below represents the average mass at each multiplet of the heavy baryons (see at the left side of the chart below). The difference between the multiplets average masses if calculated numerically from the chart of the heavy baryons points to ~140 (Mev) as can be seen here:

1385-1232= 153 (Mev)
1530-1385=145 (Mev)
1672-1530=142 (Mev)

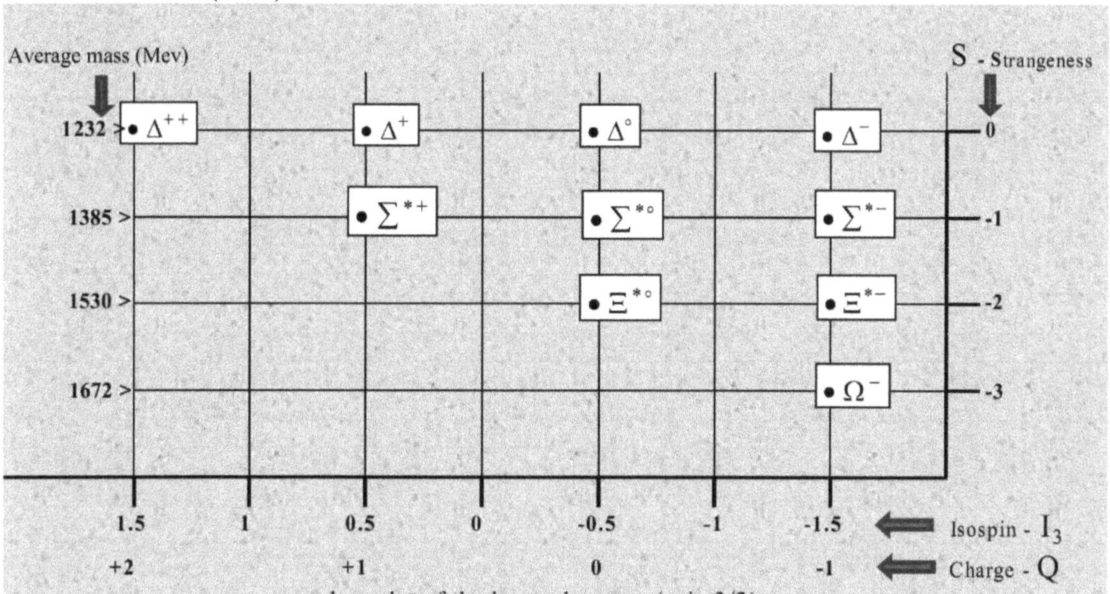

decuplet of the heavy baryons (spin 3/2)

Now let us check this difference from the equalities by means of the quarks content in the particles. We will subtract the particles with the smaller mass from the larger one to get the absolute value!

$$\Sigma^{*+} - \Delta^{+} \rightarrow uus - uud = uudd\bar{d} - uud = d\bar{d} \rightarrow \pi^{\circ}$$

$$\Xi^{*\circ} - \Sigma^{*\circ} \rightarrow uss - uds = udd\bar{d}dd\bar{d} - uddd\bar{d} = d\bar{d} \rightarrow \pi^{\circ}$$

$$\Omega^{-} - \Xi^{*-} \rightarrow sss - dss = dd\bar{d}dd\bar{d}dd\bar{d} - ddd\bar{d}dd\bar{d} = d\bar{d} \rightarrow \pi^{\circ}$$

We see that the difference between the particle's content of quarks at the relevant multiplet which indirectly reflects the average mass differences, reveals that it is equal to a mass of a π° meson which equals to ~140 Mev. This outcome can be expressed by a rule.

***The rule**: Particle from one multiplet (has the same Strangeness) is transformed into a new particle at the next multiplet that has the same Charge and Isospin I_3 by absorbing a π° meson through the strong force. It creates a composition with a new binding energy set by a new inner arrangement. Some particles have more than one option. The new interpretation of strange and charm quarks and bottom quark are used. This rule corresponds with super multiplets that include baryons or mesons (in decuplets, octets …).

For the Heavy Baryons, Heavy charmed Baryons and Heavy Bottom Baryons (spin 3/2):

$$\Delta^{-} + \pi^{\circ} \rightarrow ddd + d\bar{d} \rightarrow dd\underbrace{dd\bar{d}}_{s} \Rightarrow \Sigma^{*-}$$

$$\Delta^{\circ} + \pi^{\circ} \rightarrow udd + d\bar{d} \rightarrow ud\underbrace{dd\bar{d}}_{s} \Rightarrow \Sigma^{*\circ} \quad or \quad \Delta^{\circ} + \pi^{\circ} \rightarrow udd + u\bar{u} \rightarrow dd\underbrace{uu\bar{u}}_{c} \Rightarrow \Sigma_c^{*\circ}$$

$$\Delta^{+} + \pi^{\circ} \rightarrow uud + d\bar{d} \rightarrow uu\underbrace{dd\bar{d}}_{s} \Rightarrow \Sigma^{*+} \quad or \quad \Delta^{+} + \pi^{\circ} \rightarrow uud + u\bar{u} \rightarrow ud\underbrace{uu\bar{u}}_{c} \Rightarrow \Sigma_c^{*+}$$

$$\Sigma^{*-} + \pi^{\circ} \rightarrow dds + d\bar{d} \rightarrow ds\underbrace{dd\bar{d}}_{s} \Rightarrow \Xi^{*-} \quad or \quad \Sigma^{*-} + \pi^{\circ} \rightarrow dds + u\bar{u} \rightarrow dd\underbrace{su\bar{u}}_{b} \Rightarrow \Sigma_{b}^{*-}$$

$$\Sigma^{*\circ} + \pi^{\circ} \rightarrow uds + d\bar{d} \rightarrow us\underbrace{dd\bar{d}}_{s} \Rightarrow \Xi^{*\circ} \quad or \quad \Sigma^{*\circ} + \pi^{\circ} \rightarrow uds + u\bar{u} \rightarrow ds\underbrace{uu\bar{u}}_{c} \Rightarrow \Xi_{c}^{*\circ}$$

$$or \quad \Sigma^{*\circ} + \pi^{\circ} \rightarrow uds + u\bar{u} \rightarrow ud\underbrace{su\bar{u}}_{b} \Rightarrow \Sigma_{b}^{*\circ}$$

$$\Sigma^{*+} + \pi^{\circ} \rightarrow uus + u\bar{u} \rightarrow us\underbrace{uu\bar{u}}_{c} \Rightarrow \Xi_{c}^{*+} \quad or \quad \Sigma^{*+} + \pi^{\circ} \rightarrow uus + u\bar{u} \rightarrow uu\underbrace{su\bar{u}}_{b} \Rightarrow \Sigma_{b}^{*+}$$

$$\Xi^{*-} + \pi^{\circ} \rightarrow dss + d\bar{d} \rightarrow ss\underbrace{dd\bar{d}}_{s} \Rightarrow \Omega^{-} \quad or \quad \Xi^{*-} + \pi^{\circ} \rightarrow dss + u\bar{u} \rightarrow ds\underbrace{su\bar{u}}_{b} \Rightarrow \Xi_{b}^{*-}$$

$$\Xi^{*\circ} + \pi^{\circ} \rightarrow uss + u\bar{u} \rightarrow ss\underbrace{uu\bar{u}}_{c} \Rightarrow \Omega_{c}^{*\circ} \quad or \quad \Xi^{*\circ} + \pi^{\circ} \rightarrow uss + u\bar{u} \rightarrow us\underbrace{su\bar{u}}_{b} \Rightarrow \Xi_{b}^{*\circ}$$

$$\Omega^{-} + \pi^{\circ} \rightarrow sss + u\bar{u} \rightarrow ss\underbrace{su\bar{u}}_{b} \Rightarrow \Omega_{b}^{*-} \quad and \quad \Sigma_{c}^{*\circ} + \pi^{\circ} \rightarrow ddc + u\bar{u} \rightarrow dc\underbrace{du\bar{u}}_{s} \Rightarrow \Xi_{c}^{*\circ}$$

$$\Xi_{c}^{*\circ} + \pi^{\circ} \rightarrow dcs + u\bar{u} \rightarrow dc\underbrace{su\bar{u}}_{b} \Rightarrow \Xi_{cb}^{*\circ} \quad and \quad \Xi_{c}^{*+} + \pi^{\circ} \rightarrow ucs + u\bar{u} \rightarrow uc\underbrace{su\bar{u}}_{b} \Rightarrow \Xi_{cb}^{*+}$$

$$\Omega_{c}^{*\circ} + \pi^{\circ} \rightarrow ssc + u\bar{u} \rightarrow sc\underbrace{su\bar{u}}_{b} \Rightarrow \Omega_{cb}^{*\circ} \quad and \quad \Omega_{cc}^{*+} + \pi^{\circ} \rightarrow ccs + u\bar{u} \rightarrow cc\underbrace{su\bar{u}}_{b} \Rightarrow \Omega_{ccb}^{*+}$$

$$\Sigma_{b}^{*-} + \pi^{\circ} \rightarrow ddb + u\bar{u} \rightarrow db\underbrace{du\bar{u}}_{s} \Rightarrow \Xi_{b}^{*-}$$

$$\Sigma_{b}^{*\circ} + \pi^{\circ} \rightarrow udb + u\bar{u} \rightarrow ub\underbrace{du\bar{u}}_{s} \Rightarrow \Xi_{b}^{*\circ} \quad or \quad \Sigma_{b}^{*\circ} + \pi^{\circ} \rightarrow udb + u\bar{u} \rightarrow \underbrace{du\bar{u}}_{c}b \Rightarrow \Xi_{cb}^{*\circ}$$

$$\Sigma_{b}^{*+} + \pi^{\circ} \rightarrow uub + u\bar{u} \rightarrow \underbrace{uu\bar{u}}_{c}b \Rightarrow \Xi_{cb}^{*+} \quad and \quad \Xi_{b}^{*-} + \pi^{\circ} \rightarrow dbs + u\bar{u} \rightarrow db\underbrace{su\bar{u}}_{b} \Rightarrow \Xi_{bb}^{*-}$$

$$\Xi_{b}^{*\circ} + \pi^{\circ} \rightarrow ubs + u\bar{u} \rightarrow ub\underbrace{su\bar{u}}_{b} \Rightarrow \Xi_{bb}^{*\circ} \quad and \quad \Omega_{b}^{*-} + \pi^{\circ} \rightarrow ssb + u\bar{u} \rightarrow sb\underbrace{su\bar{u}}_{b} \Rightarrow \Omega_{bb}^{*-}$$

$$\Xi_{cb}^{*\circ} + \pi^{\circ} \rightarrow dcb + u\bar{u} \rightarrow \underbrace{du\bar{u}}_{s}cb \Rightarrow \Omega_{cb}^{*\circ} \quad and \quad \Xi_{cb}^{*+} + \pi^{\circ} \rightarrow ucb + u\bar{u} \rightarrow \underbrace{uu\bar{u}}_{c}cb \Rightarrow \Omega_{ccb}^{*+}$$

$$\Omega_{cb}^{*\circ} + \pi^{\circ} \rightarrow cbs + u\bar{u} \rightarrow cb\underbrace{su\bar{u}}_{b} \Rightarrow \Omega_{cbb}^{*\circ} \quad and \quad \Omega_{bb}^{*-} + \pi^{\circ} \rightarrow bbs + u\bar{u} \rightarrow bb\underbrace{su\bar{u}}_{b} \Rightarrow \Omega_{bbb}^{*-}$$

For the Semi Stable Baryons, Charmed Baryons and bottom baryons (spin 1/2):

$$\mathrm{p} + \pi^{\circ} \rightarrow uud + d\bar{d} \rightarrow uu\underbrace{dd\bar{d}}_{s} \Rightarrow \Sigma^{+} \quad or \quad \mathrm{p} + \pi^{\circ} \rightarrow uud + u\bar{u} \rightarrow ud\underbrace{uu\bar{u}}_{c} \Rightarrow \Sigma_{c}^{+}$$

$$\mathrm{n} + \pi^{\circ} \rightarrow udd + d\bar{d} \rightarrow ud\underbrace{dd\bar{d}}_{s} \Rightarrow \Sigma^{\circ} \quad or \quad \mathrm{n} + \pi^{\circ} \rightarrow udd + u\bar{u} \rightarrow dd\underbrace{uu\bar{u}}_{c} \Rightarrow \Sigma_{c}^{\circ}$$

$$\Sigma^{-} + \pi^{\circ} \rightarrow dds + d\bar{d} \rightarrow ds\underbrace{dd\bar{d}}_{s} \Rightarrow \Xi^{-} \quad or \quad \Sigma^{-} + \pi^{\circ} \rightarrow dds + u\bar{u} \rightarrow dd\underbrace{su\bar{u}}_{b} \Rightarrow \Sigma_{b}^{-}$$

$$\Sigma^{\circ} + \pi^{\circ} \rightarrow uds + d\bar{d} \rightarrow us\underbrace{dd\bar{d}}_{s} \Rightarrow \Xi^{\circ} \quad or \quad \Sigma^{\circ} + \pi^{\circ} \rightarrow uds + u\bar{u} \rightarrow ds\underbrace{uu\bar{u}}_{c} \Rightarrow \Xi_{c}^{\circ}$$

$$or \quad \Sigma^{\circ} + \pi^{\circ} \rightarrow uds + u\bar{u} \rightarrow ud\underbrace{su\bar{u}}_{b} \Rightarrow \Sigma_{b}^{\circ}$$

$\Xi^- + \pi^\circ \rightarrow dss + u\bar{u} \rightarrow ds\underbrace{su\bar{u}}_{b} \Rightarrow \Xi_b^-$

$\Xi^\circ + \pi^\circ \rightarrow uss + u\bar{u} \rightarrow us\underbrace{su\bar{u}}_{b} \Rightarrow \Xi_b^\circ$ *and* $\Lambda^\circ + \pi^\circ \rightarrow uds + d\bar{d} \rightarrow us\underbrace{dd\bar{d}}_{s} \Rightarrow \Xi^\circ$

$\Sigma_c^+ + \pi^\circ \rightarrow udc + u\bar{u} \rightarrow u\underbrace{du\bar{u}}_{s}c \Rightarrow \Xi_c^+$ *or* $\Sigma_c^+ + \pi^\circ \rightarrow udc + u\bar{u} \rightarrow dc\underbrace{uu\bar{u}}_{c} \Rightarrow \Xi_{cc}^+$

$\Sigma_c^{++} + \pi^\circ \rightarrow uuc + u\bar{u} \rightarrow \underbrace{uu\bar{u}}_{c}c \Rightarrow \Xi_{cc}^{++}$ *and* $\Sigma_c^\circ + \pi^\circ \rightarrow ddc + u\bar{u} \rightarrow dc\underbrace{du\bar{u}}_{s} \Rightarrow \Xi_c^\circ$

$\Xi_c^\circ + \pi^\circ \rightarrow dsc + u\bar{u} \rightarrow \underbrace{du\bar{u}}_{s}sc \Rightarrow \Omega_c^\circ$ *or* $\Xi_c^\circ + \pi^\circ \rightarrow dsc + u\bar{u} \rightarrow dc\underbrace{su\bar{u}}_{b} \Rightarrow \Xi_{cb}^\circ$

$\Xi_c^+ + \pi^\circ \rightarrow usc + u\bar{u} \rightarrow sc\underbrace{uu\bar{u}}_{c} \Rightarrow \Omega_{cc}^+$ *or* $\Xi_c^+ + \pi^\circ \rightarrow usc + u\bar{u} \rightarrow uc\underbrace{su\bar{u}}_{b} \Rightarrow \Xi_{cb}^+$

$\Xi_{cc}^+ + \pi^\circ \rightarrow dcc + u\bar{u} \rightarrow \underbrace{du\bar{u}}_{s}cc \Rightarrow \Omega_{cc}^+$ *and* $\Omega_c^\circ + \pi^\circ \rightarrow ssc + u\bar{u} \rightarrow sc\underbrace{su\bar{u}}_{b} \Rightarrow \Omega_{cb}^\circ$

$\Sigma_b^\circ + \pi^\circ \rightarrow udb + u\bar{u} \rightarrow ub\underbrace{du\bar{u}}_{s} \Rightarrow \Xi_b^\circ$ *or* $\Sigma_b^\circ + \pi^\circ \rightarrow udb + u\bar{u} \rightarrow d\underbrace{uu\bar{u}}_{c}b \Rightarrow \Xi_{cb}^\circ$

$\Sigma_b^- + \pi^\circ \rightarrow ddb + d\bar{d} \rightarrow db\underbrace{dd\bar{d}}_{s} \Rightarrow \Xi_b^-$ *and* $\Sigma_b^+ + \pi^\circ \rightarrow uub + u\bar{u} \rightarrow \underbrace{uu\bar{u}}_{c}b \Rightarrow \Xi_{cb}^+$

$\Xi_b^- + \pi^\circ \rightarrow dsb + u\bar{u} \rightarrow sb\underbrace{du\bar{u}}_{s} \Rightarrow \Omega_b^-$ *or* $\Xi_b^- + \pi^\circ \rightarrow dsb + u\bar{u} \rightarrow db\underbrace{su\bar{u}}_{b} \Rightarrow \Xi_{bb}^-$

$\Xi_b^\circ + \pi^\circ \rightarrow usb + u\bar{u} \rightarrow s\underbrace{uu\bar{u}}_{c}b \Rightarrow \Omega_{cb}^\circ$ *or* $\Xi_b^\circ + \pi^\circ \rightarrow usb + u\bar{u} \rightarrow ub\underbrace{su\bar{u}}_{b} \Rightarrow \Xi_{bb}^\circ$

$\Omega_b^- + \pi^\circ \rightarrow ssb + u\bar{u} \rightarrow sb\underbrace{su\bar{u}}_{b} \Rightarrow \Omega_{bb}^-$ *and* $\Xi_{bb}^- + \pi^\circ \rightarrow dbb + u\bar{u} \rightarrow \underbrace{du\bar{u}}_{s}bb \Rightarrow \Omega_{bb}^-$

$\Xi_{cb}^\circ + \pi^\circ \rightarrow dcb + u\bar{u} \rightarrow \underbrace{du\bar{u}}_{s}cb \Rightarrow \Omega_{cb}^\circ$ *and* $\Xi_{bb}^\circ + \pi^\circ \rightarrow ubb + u\bar{u} \rightarrow \underbrace{uu\bar{u}}_{c}bb \Rightarrow \Omega_{cbb}^\circ$

$\Xi_{cb}^+ + \pi^\circ \rightarrow ucb + u\bar{u} \rightarrow \underbrace{uu\bar{u}}_{c}cb \Rightarrow \Omega_{ccb}^+$ *and* $\Omega_{cb}^\circ + \pi^\circ \rightarrow scb + u\bar{u} \rightarrow cb\underbrace{su\bar{u}}_{b} \Rightarrow \Omega_{cbb}^\circ$

For the Semi Stable Mesons and Charmed Mesons:

$\pi^- + \pi^\circ \rightarrow \bar{u}d + d\bar{d} \rightarrow \bar{u}\underbrace{dd\bar{d}}_{s} \Rightarrow K^-$ *and* $\pi^+ + \pi^\circ \rightarrow u\bar{d} + d\bar{d} \rightarrow u\underbrace{\bar{d}d\bar{d}}_{\bar{s}} \Rightarrow K^+$

$\pi^\circ + \pi^\circ \rightarrow d\bar{d} + d\bar{d} \rightarrow d\underbrace{\bar{d}d\bar{d}}_{\bar{s}} \Rightarrow K^\circ$ *or* $\pi^\circ + \pi^\circ \rightarrow d\bar{d} + d\bar{d} \rightarrow \underbrace{d\bar{d}d}_{s}\bar{d} \Rightarrow \bar{K}^\circ$

$\pi^\circ + \pi^\circ \rightarrow u\bar{u} + u\bar{u} \rightarrow \underbrace{u\bar{u}u}_{c}\bar{u} \Rightarrow D^\circ$ *or* $\pi^\circ + \pi^\circ \rightarrow u\bar{u} + u\bar{u} \rightarrow u\underbrace{\bar{u}u\bar{u}}_{\bar{c}} \Rightarrow \bar{D}^\circ$

$\pi^- + \pi^\circ \rightarrow \bar{u}d + u\bar{u} \rightarrow d\underbrace{\bar{u}u\bar{u}}_{\bar{c}} \Rightarrow D^-$ *and* $\pi^+ + \pi^\circ \rightarrow u\bar{d} + u\bar{u} \rightarrow \bar{d}\underbrace{uu\bar{u}}_{c} \Rightarrow D^+$

More examples presented schematically shows how particles are transformed into new ones by absorbing one π° or one π^+ thru strong force interactions. It creates a composition with new bind energy set by a new inner arrangement. Some particles have more than one option. The new interpretation of strange, charm and bottom quarks presented previously are used. The arrows indicate the outcome of each composition!

Heavy Baryons (spin 3/2)

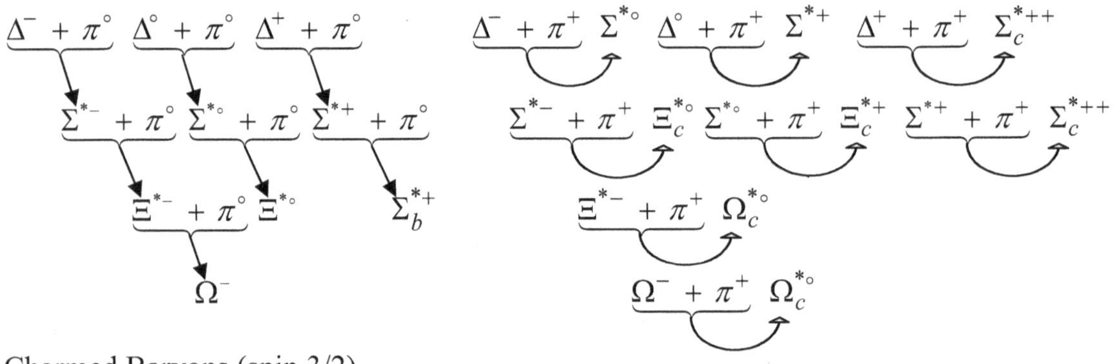

$$\underbrace{\Delta^- + \pi^\circ} \qquad \underbrace{\Delta^\circ + \pi^\circ} \qquad \underbrace{\Delta^+ + \pi^\circ} \qquad\qquad \underbrace{\Delta^- + \pi^+}\;\Sigma^{*\circ} \qquad \underbrace{\Delta^\circ + \pi^+}\;\Sigma^{*+} \qquad \underbrace{\Delta^+ + \pi^+}\;\Sigma_c^{*++}$$

$$\underbrace{\Sigma^{*-} + \pi^\circ} \quad \underbrace{\Sigma^{*\circ} + \pi^\circ} \quad \underbrace{\Sigma^{*+} + \pi^\circ} \qquad \underbrace{\Sigma^{*-} + \pi^+}\;\Xi_c^{*\circ} \quad \underbrace{\Sigma^{*\circ} + \pi^+}\;\Xi_c^{*+} \quad \underbrace{\Sigma^{*+} + \pi^+}\;\Sigma_c^{*++}$$

$$\underbrace{\Xi^{*-} + \pi^\circ}\;\Xi^{*\circ} \qquad \Sigma_b^{*+} \qquad\qquad \underbrace{\Xi^{*-} + \pi^+}\;\Omega_c^{*\circ}$$

$$\Omega^- \qquad\qquad\qquad \underbrace{\Omega^- + \pi^+}\;\Omega_c^{*\circ}$$

Charmed Baryons (spin 3/2)

$$\underbrace{\Sigma_c^{*\circ} + \pi^\circ} \qquad \underbrace{\Sigma_c^{*+} + \pi^\circ} \qquad\qquad \underbrace{\Sigma_c^{*\circ} + \pi^+}\;\Xi_{cc}^{*+} \qquad \underbrace{\Sigma_c^{*+} + \pi^+}\;\Xi_{cc}^{*++}$$

$$\Xi_c^{*\circ} \qquad\qquad \Xi_c^{*+} \qquad\qquad \underbrace{\Xi_c^{*\circ} + \pi^+}\;\Omega_{cc}^+ \quad \underbrace{\Xi_c^{*+} + \pi^+}\;\Xi_{cc}^{*++} \quad \underbrace{\Xi_{cc}^{*+} + \pi^+}\;\Omega_{ccc}^{++}$$

$$\underbrace{\Xi_c^{*\circ} + \pi^\circ} \qquad \underbrace{\Xi_{cc}^{*+} + \pi^\circ}$$

$$\Omega_c^{*\circ} \qquad\qquad \Omega_{cc}^{*+} \qquad \underbrace{\Omega_c^{*\circ} + \pi^+}\;\Omega_{cc}^{*+} \qquad \underbrace{\Omega_{cc}^{*+} + \pi^+}\;\Omega_{ccc}^{++}$$

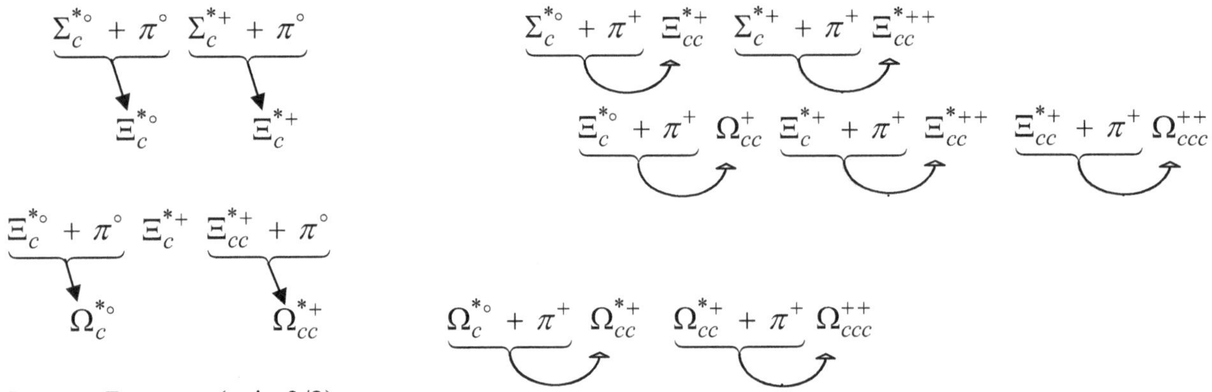

Bottom Baryons (spin 3/2)

$$\underbrace{\Sigma_b^{*-} + \pi^\circ} \quad \underbrace{\Sigma_b^{*\circ} + \pi^\circ} \qquad\qquad \underbrace{\Sigma_b^{*-} + \pi^+}\;\Xi_{cb}^{*\circ} \quad \underbrace{\Sigma_b^{*\circ} + \pi^+}\;\Xi_{cb}^{*+}$$

$$\underbrace{\Xi_b^{*-} + \pi^\circ} \quad \underbrace{\Xi_b^{*\circ} + \pi^\circ} \quad \underbrace{\Xi_{cb}^{*\circ} + \pi^\circ} \quad \underbrace{\Xi_{bb}^{*-} + \pi^\circ} \quad \underbrace{\Xi_b^{*-} + \pi^+}\;\Omega_{cb}^\circ \quad \underbrace{\Xi_b^{*\circ} + \pi^+}\;\Xi_{cb}^{*+} \quad \underbrace{\Xi_{cb}^{*\circ} + \pi^+}\;\Omega_{ccb}^{*+}$$

$$\Omega_b^{*-} \qquad\qquad \Omega_{cb}^{*\circ} \qquad\qquad \Omega_{bb}^{*-} \qquad \underbrace{\Omega_b^{*-} + \pi^+}\;\Omega_{cb}^{*\circ} \quad \underbrace{\Omega_{cb}^{*\circ} + \pi^+}\;\Omega_{ccb}^{*+} \quad \underbrace{\Omega_{bb}^{*-} + \pi^+}\;\Omega_{cbb}^{*\circ}$$

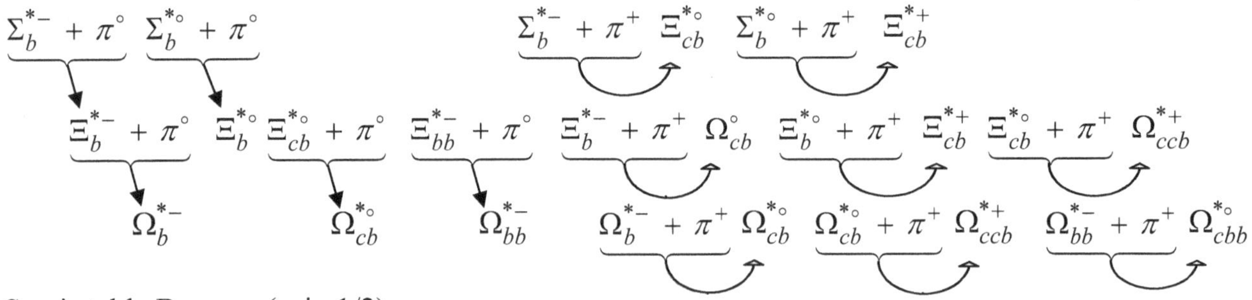

Semi stable Baryons (spin 1/2)

$$\underbrace{n + \pi^\circ} \qquad \underbrace{p + \pi^\circ} \qquad\qquad \underbrace{n + \pi^+}\;\Sigma_c^+ \qquad \underbrace{p + \pi^+}\;\Sigma_c^{++}$$

$$\underbrace{\Sigma^- + \pi^\circ} \quad \underbrace{\Sigma^\circ + \pi^\circ}\;\Sigma^+ \qquad \underbrace{\Sigma^- + \pi^+}\;\Xi_c^\circ \;\;\text{Or}\;\; \underbrace{\Sigma^- + \pi^+}\;\Sigma_c^\circ \quad \underbrace{\Sigma^\circ + \pi^+}\;\Xi_c^+ \quad \underbrace{\Sigma^+ + \pi^+}\;\Sigma_c^{++}$$

$$\Xi^- \qquad\qquad \Xi^\circ \qquad\qquad \underbrace{\Xi^- + \pi^+}\;\Omega_c^\circ \quad \underbrace{\Xi^\circ + \pi^+}\;\Xi_c^+$$

Charmed Baryons (spin 1/2)

$$\underbrace{\Sigma_c^\circ + \pi^\circ} \quad \underbrace{\Sigma_c^+ + \pi^\circ} \qquad\qquad \underbrace{\Sigma_c^\circ + \pi^+}\;\Xi_{cc}^+ \quad \underbrace{\Sigma_c^+ + \pi^+}\;\Xi_{cc}^{++}$$

$$\underbrace{\Xi_c^\circ + \pi^\circ}\;\Xi_c^+ \qquad\qquad \underbrace{\Xi_c^\circ + \pi^+}\;\Omega_{cc}^+ \quad \underbrace{\Xi_c^+ + \pi^+}\;\Xi_{cc}^{++}$$

$$\Omega_c^\circ$$

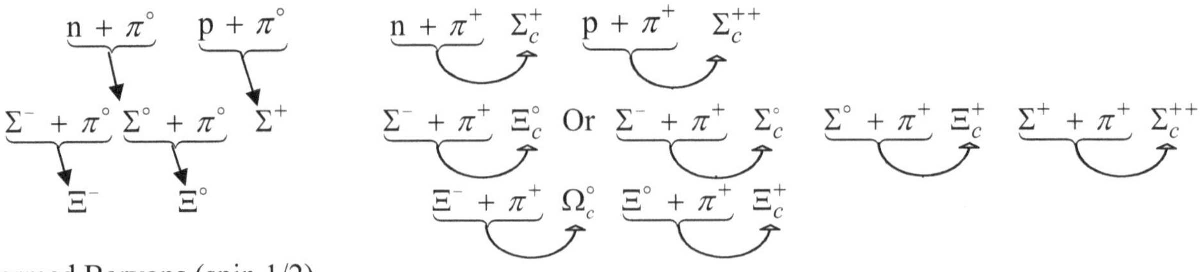

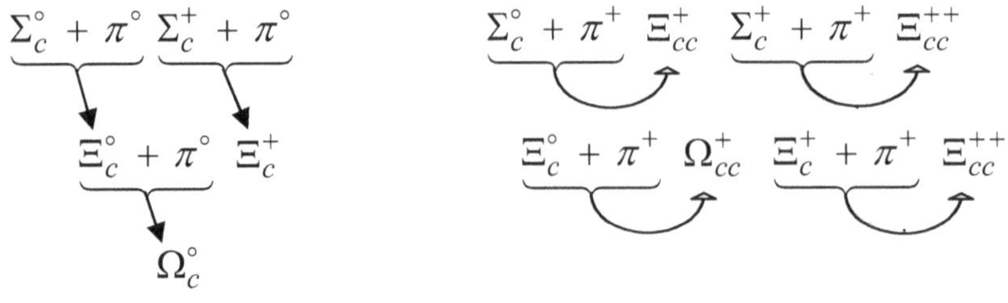

Bottom Baryons (spin 1/2)

$$\underbrace{\Sigma_b^- + \pi^\circ}\ \underbrace{\Sigma_b^\circ + \pi^\circ} \qquad\qquad \underbrace{\Sigma_b^- + \pi^+}\ \underbrace{\Xi_{cb}^\circ + \pi^+}\ \Omega_{ccb}^+ \quad \underbrace{\Sigma_b^\circ + \pi^+}\ \Xi_{cb}^+$$

$$\underbrace{\Xi_b^- + \pi^\circ}\ \underbrace{\Xi_b^\circ + \pi^\circ}\ \underbrace{\Xi_{cb}^\circ + \pi^\circ}\ \underbrace{\Xi_{bb}^- + \pi^\circ} \quad \underbrace{\Xi_b^- + \pi^+}\ \Omega_{cb}^\circ \quad \underbrace{\Xi_b^\circ + \pi^+}\ \Xi_{cb}^+ \quad \underbrace{\Xi_{bb}^- + \pi^+}\ \Omega_{cbb}^\circ$$

$$\Omega_b^- \qquad\qquad \Omega_{cb}^\circ \qquad\qquad \Omega_{bb}^- \qquad \underbrace{\Omega_b^- + \pi^+}\ \Omega_{cb}^\circ \quad \underbrace{\Omega_{cb}^\circ + \pi^+}\ \Omega_{ccb}^+ \quad \underbrace{\Omega_{bb}^- + \pi^+}\ \Omega_{cbb}^\circ$$

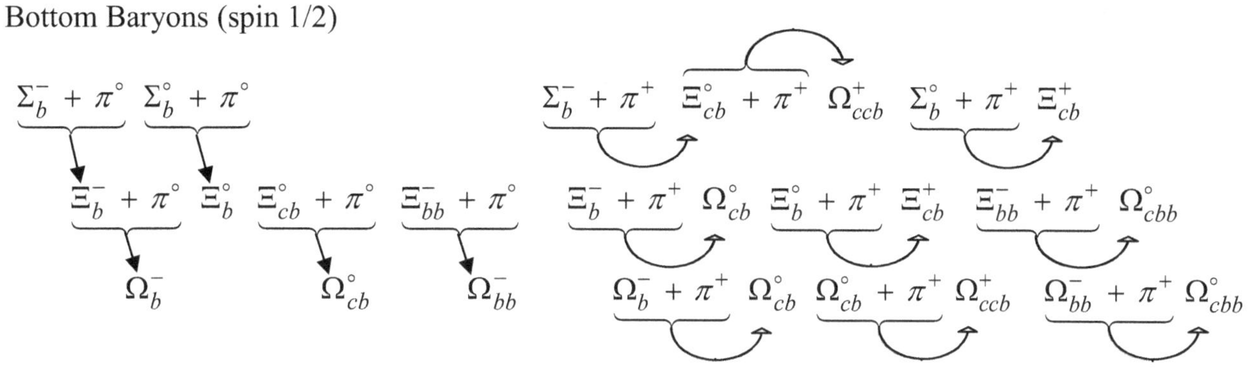

* Now a few examples that deal with the quarks spin orientation within the sub atomic particles!

a. The case of the missing proton spin "the proton spin crisis" resolution:

In chapter 12 (The quarks) I have noted in paragraph (c) the following clarification:
Because each quark possesses one-third of the proton or the neutron mass it suggests that the orbital angular momentum $\vec{L}_{quark}$ of the quarks within the baryons is a factor of thirds of $\hbar$ which means as can be seen in the schema below that for the first level it is $(1/3\hbar)$ and for the second level $(2/3\hbar)$.

I must note here that this result differs from the quark's own spin $(1/2\hbar)$! In addition the $\vec{L}_{quark}$ can be either negative or positive depending on the quark's revolving direction at the relevant orbital. For the quarks at the first orbital $\vec{L}_{quark}$ it is negative and for the second it is positive, and for the anti quarks at the first orbital $\vec{L}_{quark}$ it is positive and for the second it is negative. There is one exception which is enforced by the Pauli Exclusion Principle saying that **no** two identical particles with spin $1/2\hbar$ can occupy the same quantum state simultaneously.

From this insight we can resolve the so called "The case of the missing proton spin". Here below is a schema which describes how the proton gets a total angular momentum j^p (spin +orbital $=+1/2\hbar$) of which was experimentally obtained. It is clearly seen that the missing spin is resolved by the contribution of the orbital angular momentum within the proton. The schema below shows two orbital levels in the proton occupied by the three quarks - uud. The arrows hold for the revolving directions! The positive spin marked as $(\uparrow)$ and the negative spin marked as $(\downarrow)$ and for the orbital angular momentum the negative is marked as $(\downarrow)$ and the positive is marked as $(\uparrow)$.

$$\mathbf{p}$$

$$\left(\begin{array}{c} \underline{spin} \qquad \underline{orbital} \\ +\frac{1}{2}\hbar \ \uparrow \ u \ \uparrow \ +\frac{2}{3}\hbar \\ \underline{spin} \qquad \underline{orbital} \\ -\frac{1}{2}\hbar \ \downarrow \ u \ \uparrow \ +\frac{2}{3}\hbar \\ \underline{spin} \qquad \underline{orbital} \\ -\frac{1}{2}\hbar \ \downarrow \ d \ \downarrow \ -\frac{1}{3}\hbar \end{array}\right)$$

Total spin: $-1/2\hbar$
Total orbital: $+1\hbar$
Total j^p (spin + orbital): $+1/2\hbar$

* Production of the Delta Baryons via the strong force interaction:

How neutron or proton are transformed into the Delta Baryons by absorbing one π^+ or one π^- thru the strong force interactions that creates an excited states compositions with a new binding energy and a spin $j^P(\text{spin} + \text{orbital}) = \pm 3/2\,\hbar$ acquired by a new inner arrangement!

1. The proton transforms into a Δ^{++} **(an excited proton)**.

The proton absorbs in a reaction a π^+ meson and transforms into a Δ^{++} (see also the schema below).

After being absorbed the π^+ meson reaches the third level in the proton and it is positioned there for a split of second (this level permits a charge of $e^{\pm}$) and it is emitted out instantly before acquiring the energy of the third level, hence it is considered as a decaying process via the strong force (the process do not include the weak boson involvement). For an outside observer it looks like the process is as following:

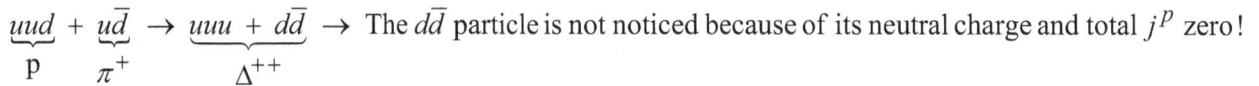

$$\underbrace{uud}_{p} + \underbrace{u\bar{d}}_{\pi^+} \rightarrow \underbrace{uuu + d\bar{d}}_{\Delta^{++}} \rightarrow \text{The } d\bar{d} \text{ particle is not noticed because of its neutral charge and total } j^P \text{ zero!}$$

The $d\bar{d}$ component is the neutral π° meson which is "invisible" (zero charge and a total $j^P = 0$) hence the Δ^{++} particle seems to consist of three *up* quarks only: *uuu*

Because the π^+ meson which is the composition of $u\bar{d}$ quarks is positioned at the third level, while the rest two *up* quarks and one *down* quark of the proton are positioned at the second and the first levels, there is no conflict with the Pauli Exclusion. The schema below shows it clearly:

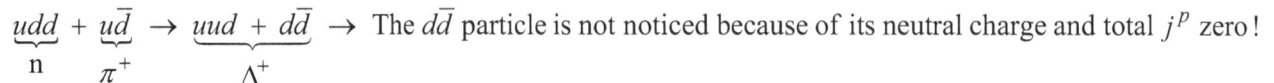

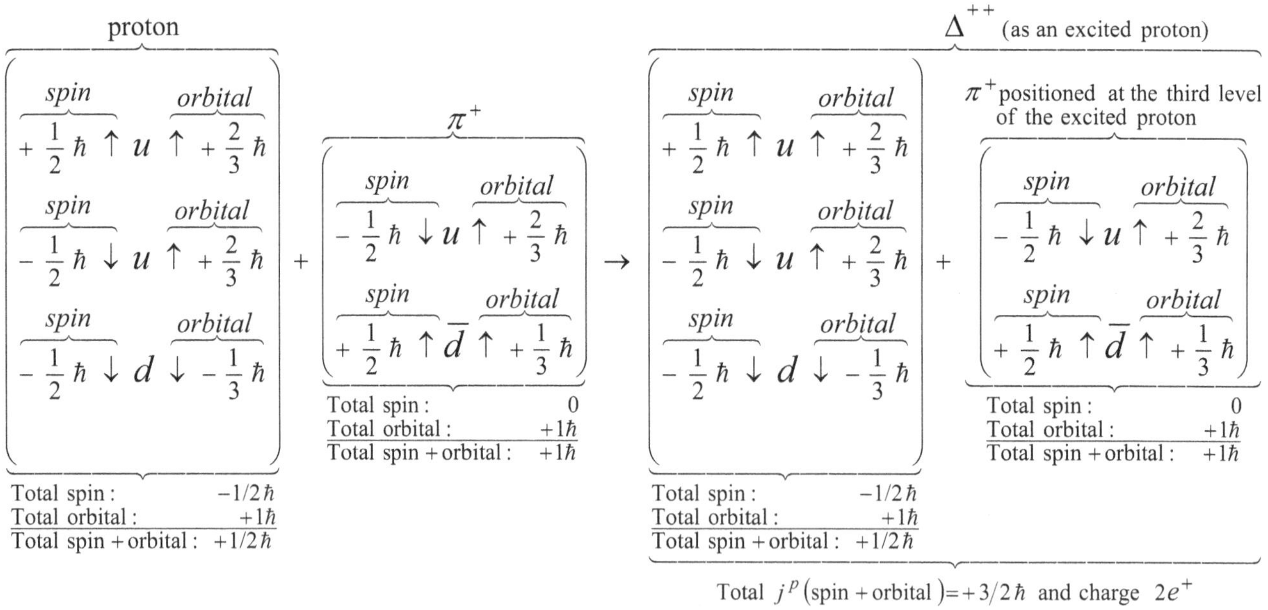

In this mode Δ^{++} decays to the original particles that created it via the strong force: $\Delta^{++} \rightarrow P + \pi^+$

2. The neutron transforms into Δ^+ (an excited neutron).

The neutron absorbs in a reaction a π^+ meson and transforms into a Δ^+ (see the schema next page):

After being absorbed the π^+ meson reaches the third level in the neutron and it is positioned there for a split of second and it is emitted out instantly before acquiring the energy of the third level, hence it is considered as a decaying process via the strong force (meaning that the process do not include the weak boson involvement). For an outside observer it looks like the process is as:

$$\underbrace{udd}_{n} + \underbrace{u\bar{d}}_{\pi^+} \rightarrow \underbrace{uud + d\bar{d}}_{\Delta^{+}} \rightarrow \text{The } d\bar{d} \text{ particle is not noticed because of its neutral charge and total } j^P \text{ zero!}$$

The $d\bar{d}$ component is the neutral π° meson which is "invisible" (zero charge and a total $j^P = 0$).

Hence the Δ^+ particle seems to consist of two *up* quarks and one *down* quark only.

Because the π^+ meson which is the composition of $u\bar{d}$ quarks is positioned at the third level while the rest of one *up* quark and two *down* quarks of the neutron are positioned at the second and the first levels respectively, there is no conflict with the Pauli Exclusion. The schema shows it clearly:

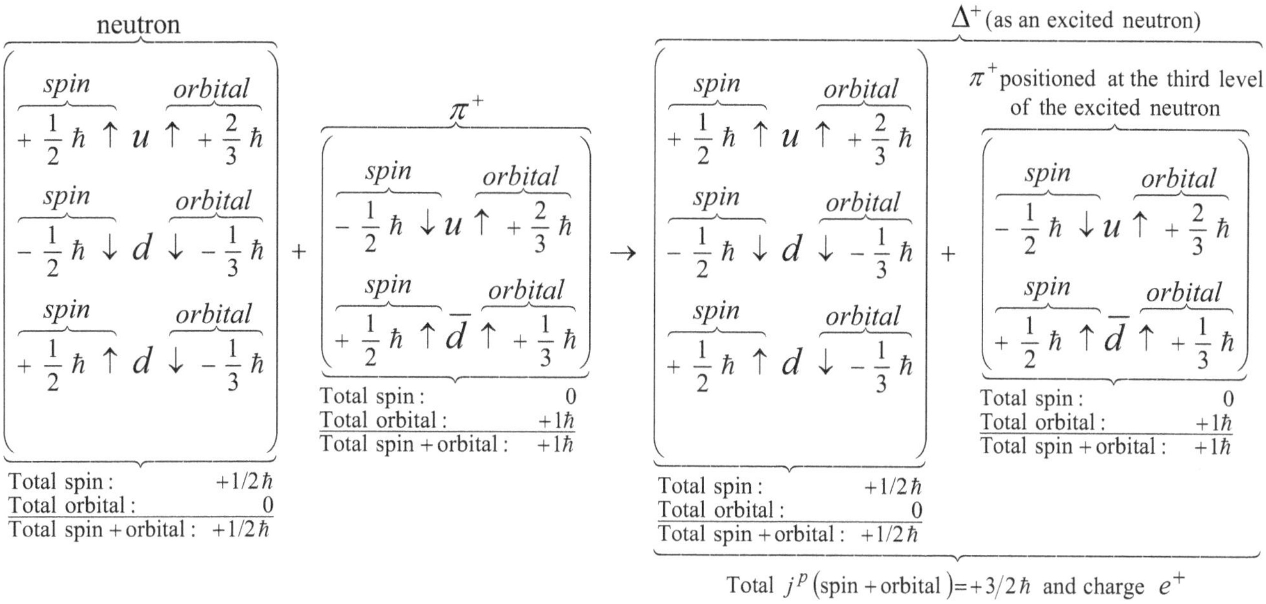

neutron

$$
\left(
\begin{array}{c}
\overbrace{+\tfrac{1}{2}\hbar\ \uparrow}^{spin}\ u\ \overbrace{\uparrow\ +\tfrac{2}{3}\hbar}^{orbital}\\[4pt]
\overbrace{-\tfrac{1}{2}\hbar\ \downarrow}^{spin}\ d\ \overbrace{\downarrow\ -\tfrac{1}{3}\hbar}^{orbital}\\[4pt]
\overbrace{+\tfrac{1}{2}\hbar\ \uparrow}^{spin}\ d\ \overbrace{\downarrow\ -\tfrac{1}{3}\hbar}^{orbital}
\end{array}
\right)
$$

Total spin : $\quad +1/2\,\hbar$
Total orbital : $\quad 0$
Total spin + orbital : $\quad +1/2\,\hbar$

$$+ \quad \pi^+ \quad \left(
\begin{array}{c}
\overbrace{-\tfrac{1}{2}\hbar\ \downarrow}^{spin}\ u\ \overbrace{\uparrow\ +\tfrac{2}{3}\hbar}^{orbital}\\[4pt]
\overbrace{+\tfrac{1}{2}\hbar\ \uparrow}^{spin}\ \bar{d}\ \overbrace{\uparrow\ +\tfrac{1}{3}\hbar}^{orbital}
\end{array}
\right)$$

Total spin : $\quad 0$
Total orbital : $\quad +1\hbar$
Total spin + orbital : $\quad +1\hbar$

$$\rightarrow$$

Δ^+ (as an excited neutron)

$$
\left(
\begin{array}{c}
\overbrace{+\tfrac{1}{2}\hbar\ \uparrow}^{spin}\ u\ \overbrace{\uparrow\ +\tfrac{2}{3}\hbar}^{orbital}\\[4pt]
\overbrace{-\tfrac{1}{2}\hbar\ \downarrow}^{spin}\ d\ \overbrace{\downarrow\ -\tfrac{1}{3}\hbar}^{orbital}\\[4pt]
\overbrace{+\tfrac{1}{2}\hbar\ \uparrow}^{spin}\ d\ \overbrace{\downarrow\ -\tfrac{1}{3}\hbar}^{orbital}
\end{array}
\right)
$$

Total spin : $\quad +1/2\,\hbar$
Total orbital : $\quad 0$
Total spin + orbital : $\quad +1/2\,\hbar$

π^+ positioned at the third level of the excited neutron

$$+ \quad \left(
\begin{array}{c}
\overbrace{-\tfrac{1}{2}\hbar\ \downarrow}^{spin}\ u\ \overbrace{\uparrow\ +\tfrac{2}{3}\hbar}^{orbital}\\[4pt]
\overbrace{+\tfrac{1}{2}\hbar\ \uparrow}^{spin}\ \bar{d}\ \overbrace{\uparrow\ +\tfrac{1}{3}\hbar}^{orbital}
\end{array}
\right)$$

Total spin : $\quad 0$
Total orbital : $\quad +1\hbar$
Total spin + orbital : $\quad +1\hbar$

Total j^p (spin + orbital) = $+3/2\,\hbar$ and charge e^+

In this mode Δ^+ decays to the original particles that created it via the strong force: $\Delta^+ \rightarrow n + \pi^+$

A different internal arrangement of Δ^+ occurs when it transforms into an excited proton. This option is not using the third level. The Δ^+ particle is an excited form of a proton that consists of a proton and a neutral π° meson as it shows in the following schema:

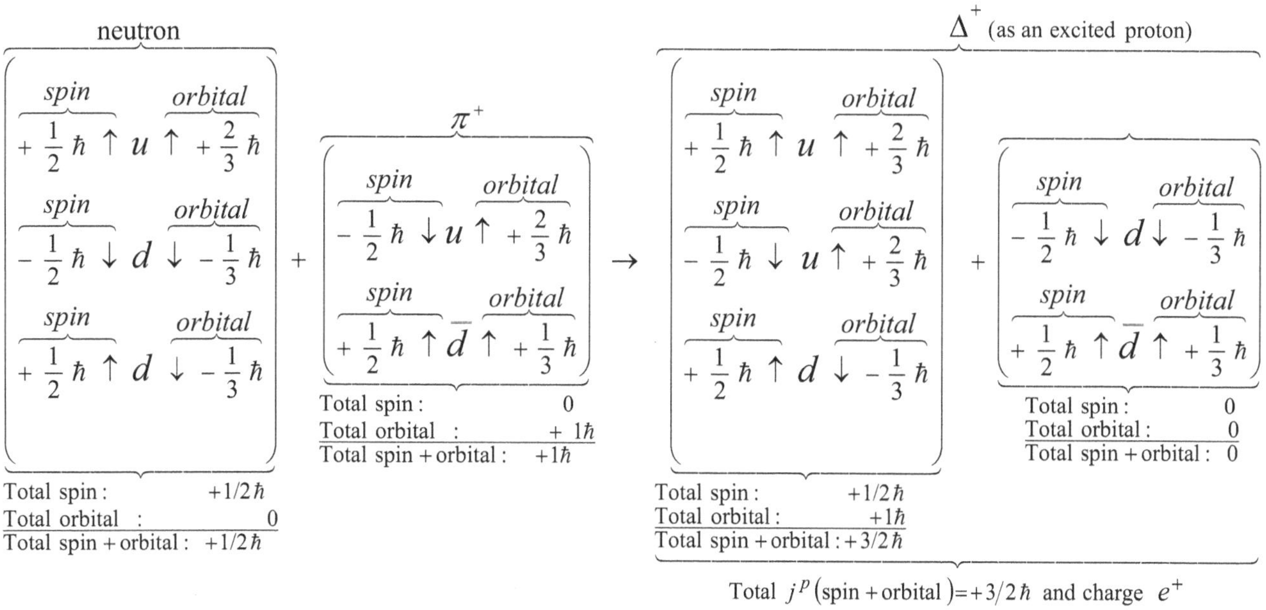

neutron

$$
\left(
\begin{array}{c}
\overbrace{+\tfrac{1}{2}\hbar\ \uparrow}^{spin}\ u\ \overbrace{\uparrow\ +\tfrac{2}{3}\hbar}^{orbital}\\[4pt]
\overbrace{-\tfrac{1}{2}\hbar\ \downarrow}^{spin}\ d\ \overbrace{\downarrow\ -\tfrac{1}{3}\hbar}^{orbital}\\[4pt]
\overbrace{+\tfrac{1}{2}\hbar\ \uparrow}^{spin}\ d\ \overbrace{\downarrow\ -\tfrac{1}{3}\hbar}^{orbital}
\end{array}
\right)
$$

Total spin : $\quad +1/2\,\hbar$
Total orbital : $\quad 0$
Total spin + orbital : $\quad +1/2\,\hbar$

$$+ \quad \pi^+ \quad \left(
\begin{array}{c}
\overbrace{-\tfrac{1}{2}\hbar\ \downarrow}^{spin}\ u\ \overbrace{\uparrow\ +\tfrac{2}{3}\hbar}^{orbital}\\[4pt]
\overbrace{+\tfrac{1}{2}\hbar\ \uparrow}^{spin}\ \bar{d}\ \overbrace{\uparrow\ +\tfrac{1}{3}\hbar}^{orbital}
\end{array}
\right)$$

Total spin : $\quad 0$
Total orbital : $\quad +1\hbar$
Total spin + orbital : $\quad +1\hbar$

$$\rightarrow$$

Δ^+ (as an excited proton)

$$
\left(
\begin{array}{c}
\overbrace{+\tfrac{1}{2}\hbar\ \uparrow}^{spin}\ u\ \overbrace{\uparrow\ +\tfrac{2}{3}\hbar}^{orbital}\\[4pt]
\overbrace{-\tfrac{1}{2}\hbar\ \downarrow}^{spin}\ u\ \overbrace{\uparrow\ +\tfrac{2}{3}\hbar}^{orbital}\\[4pt]
\overbrace{+\tfrac{1}{2}\hbar\ \uparrow}^{spin}\ d\ \overbrace{\downarrow\ -\tfrac{1}{3}\hbar}^{orbital}
\end{array}
\right)
$$

Total spin : $\quad +1/2\,\hbar$
Total orbital : $\quad +1\hbar$
Total spin + orbital : $\quad +3/2\,\hbar$

$$+ \quad \left(
\begin{array}{c}
\overbrace{-\tfrac{1}{2}\hbar\ \downarrow}^{spin}\ d\ \overbrace{\downarrow\ -\tfrac{1}{3}\hbar}^{orbital}\\[4pt]
\overbrace{+\tfrac{1}{2}\hbar\ \uparrow}^{spin}\ \bar{d}\ \overbrace{\uparrow\ +\tfrac{1}{3}\hbar}^{orbital}
\end{array}
\right)$$

Total spin : $\quad 0$
Total orbital : $\quad 0$
Total spin + orbital : $\quad 0$

Total j^p (spin + orbital) = $+3/2\,\hbar$ and charge e^+

In this mode Δ^+ decays into a proton and a π° meson via the strong force: $\Delta^+ \rightarrow p + \pi^\circ$

Please notice that in this different internal arrangement of Δ^+ there is no conflict with the Pauli Exclusion Principle. The schema above shows it clearly:

This option requires spin compensation after the π° meson is breaking away from the Δ^+ composition during the decay process! It occurs spontaneously in the proton which is part of this particle by a **spin flip** of the *down* quark from $(+1/2\,\hbar\ \uparrow)$ state to $(-1/2\,\hbar\ \downarrow)$ to form the proton's spin ($j^p = +1/2\,\hbar\ \uparrow$).

3. The proton transforms into a Δ° (an excited proton).

The proton absorbs in a reaction π^{-} (as tensor meson) and transforms into Δ° :

The π^{-} meson decays via the strong force. For an outside observer the process looks like as:

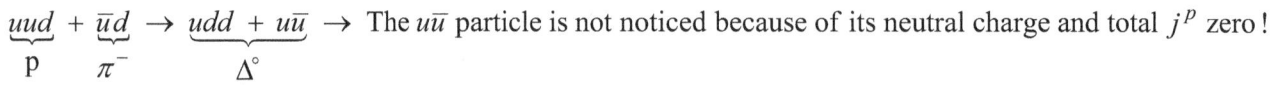

$\underbrace{uud}_{p} + \underbrace{\bar{u}d}_{\pi^{-}} \rightarrow \underbrace{udd + u\bar{u}}_{\Delta^{\circ}} \rightarrow$ The $u\bar{u}$ particle is not noticed because of its neutral charge and total j^{P} zero!

The $u\bar{u}$ component is the neutral π° meson which is "invisible" (zero charge and a total $j^{P} = 0$).

Hence the Δ° particle seems to consist of one *up* quark and two down quarks only: *udd*

Because the π^{-} meson which is the composition of $\bar{u}d$ quarks is positioned at the third level while the rest of two *up* quarks and one *down* quark of the proton are positioned at the second and the first levels respectively, there is no conflict with the Pauli Exclusion. The schema below shows it clearly:

proton

spin	quark	orbital
$+\frac{1}{2}\hbar \uparrow$	u	$\uparrow +\frac{2}{3}\hbar$
$-\frac{1}{2}\hbar \downarrow$	u	$\uparrow +\frac{2}{3}\hbar$
$-\frac{1}{2}\hbar \downarrow$	d	$\downarrow -\frac{1}{3}\hbar$

Total spin: $-1/2\hbar$
Total orbital: $+1\hbar$
Total spin + orbital: $+1/2\hbar$

+ π^{-}

spin	quark	orbital
$-\frac{1}{2}\hbar \downarrow$	$\bar{u}$	$\downarrow -\frac{2}{3}\hbar$
$-\frac{1}{2}\hbar \downarrow$	d	$\downarrow -\frac{1}{3}\hbar$

Total spin: $-1\hbar$
Total orbital: $-1\hbar$
Total spin + orbital: $-2\hbar$

$\rightarrow$ **Δ° (as an excited proton)**

spin	quark	orbital
$+\frac{1}{2}\hbar \uparrow$	u	$\uparrow +\frac{2}{3}\hbar$
$-\frac{1}{2}\hbar \downarrow$	u	$\uparrow +\frac{2}{3}\hbar$
$-\frac{1}{2}\hbar \downarrow$	d	$\downarrow -\frac{1}{3}\hbar$

Total spin: $-1/2\hbar$
Total orbital: $+1\hbar$
Total spin + orbital: $+1/2\hbar$

+ π^{-} positioned at the third level of the excited proton

spin	quark	orbital
$-\frac{1}{2}\hbar \downarrow$	$\bar{u}$	$\downarrow -\frac{2}{3}\hbar$
$-\frac{1}{2}\hbar \downarrow$	d	$\downarrow -\frac{1}{3}\hbar$

Total spin: $-1\hbar$
Total orbital: $-1\hbar$
Total spin + orbital: $-2\hbar$

Total j^{P} (spin + orbital) $= -3/2\hbar$ and Neutral charge

In this mode Δ° decays to the original particles that created it via the strong force: $\Delta^{\circ} \rightarrow p + \pi^{-}$

A different internal arrangement of Δ° occurs when it transforms into an excited neutron as:

proton

spin	quark	orbital
$+\frac{1}{2}\hbar \uparrow$	u	$\uparrow +\frac{2}{3}\hbar$
$-\frac{1}{2}\hbar \downarrow$	u	$\uparrow +\frac{2}{3}\hbar$
$-\frac{1}{2}\hbar \downarrow$	d	$\downarrow -\frac{1}{3}\hbar$

Total spin: $-1/2\hbar$
Total orbital: $+1\hbar$
Total spin + orbital: $+1/2\hbar$

+ π^{-}

spin	quark	orbital
$-\frac{1}{2}\hbar \downarrow$	$\bar{u}$	$\downarrow -\frac{2}{3}\hbar$
$-\frac{1}{2}\hbar \downarrow$	d	$\downarrow -\frac{1}{3}\hbar$

Total spin: $-1\hbar$
Total orbital: $-1\hbar$
Total spin + orbital: $-2\hbar$

$\rightarrow$ **Δ° (as an excited neutron)**

spin	quark	orbital
$-\frac{1}{2}\hbar \downarrow$	u	$\uparrow +\frac{2}{3}\hbar$
$-\frac{1}{2}\hbar \downarrow$	d	$\downarrow -\frac{1}{3}\hbar$
$-\frac{1}{2}\hbar \downarrow$	d	$\downarrow -\frac{1}{3}\hbar$

Total spin: $-3/2\hbar$
Total orbital: 0
Total spin + orbital: $-3/2\hbar$

+

spin	quark	orbital
$-\frac{1}{2}\hbar \downarrow$	$\bar{u}$	$\downarrow -\frac{2}{3}\hbar$
$+\frac{1}{2}\hbar \uparrow$	u	$\uparrow +\frac{2}{3}\hbar$

Total spin: 0
Total orbital: 0
Total spin + orbital: 0

Total j^{P} (spin + orbital) $= -3/2\hbar$ and Neutral charge

In this mode Δ° decays to a neutron and a neutral π° meson via the strong force: $\Delta^{\circ} \rightarrow n + \pi^{\circ}$.

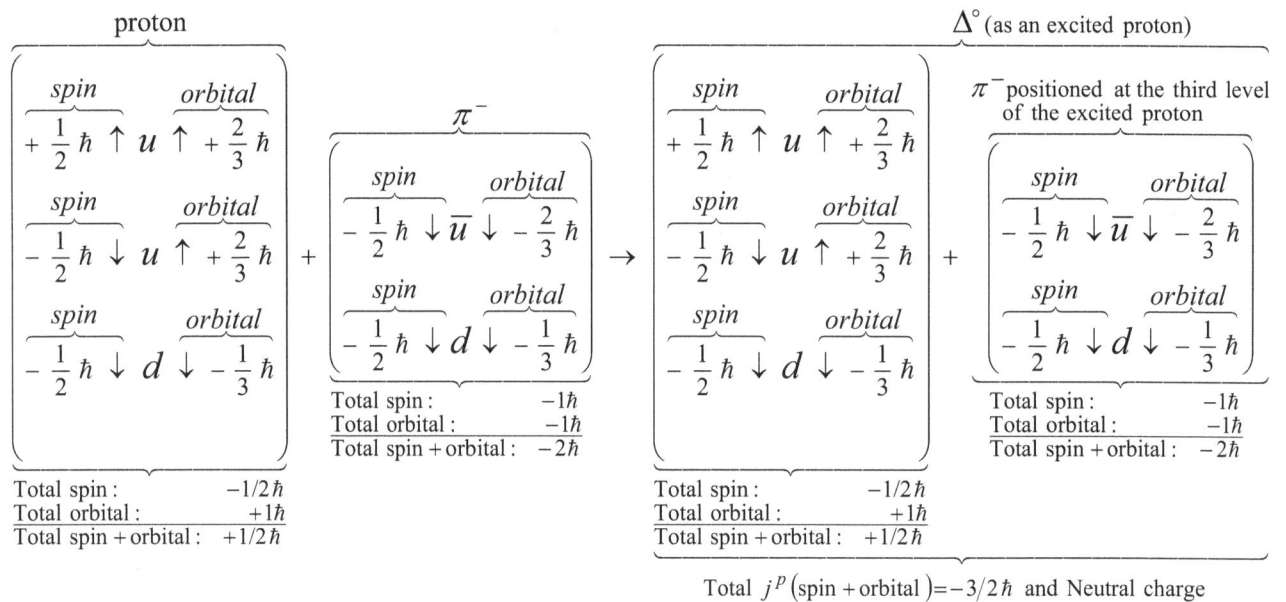

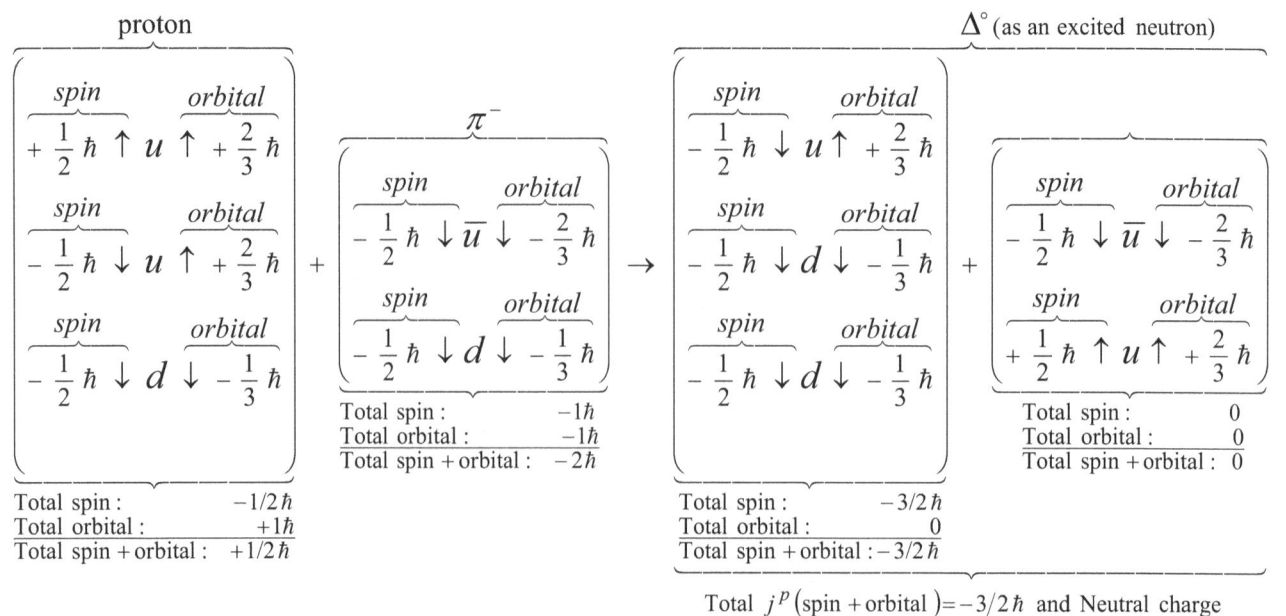

Please notice that in this different internal arrangement of Δ° there is no conflict with the Pauli Exclusion Principle.

This option requires spin compensation after the π° meson is breaking away from Δ° composition during the decay process! It occurs spontaneously within the neutron by a **spin flip** of a *down* quark from $(-\frac{1}{2}\hbar\downarrow)$ state to $(+\frac{1}{2}\hbar\uparrow)$ and the *up* quark from $(-\frac{1}{2}\hbar\downarrow)$ state to $(+\frac{1}{2}\hbar\uparrow)$ to form the neutron's spin ($j^{P} = +\frac{1}{2}\hbar\uparrow$).

4. The neutron transforms into Δ^{-} (an excited neutron).

The neutron absorbs a π^{-} (as tensor meson) and transforms into Δ^{-} :

After being absorbed π^{-} meson reaches the third level in the neutron and it is positioned there for a split of second (this level permits a charge of $e^{\pm}$) and it's emitted out instantly before acquiring the energy of the third level, hence it is considered as a decaying process via the strong force. For an outside observer it looks like the process is as following:

$\underbrace{udd}_{n} + \underbrace{\bar{u}d}_{\pi^{-}} \rightarrow \underbrace{ddd + u\bar{u}}_{\Delta^{-}} \rightarrow$ The $u\bar{u}$ particle is not noticed because of its neutral charge and total j^{P} zero!

The $u\bar{u}$ component is the neutral π° meson which is "invisible" (zero charge and a total $j^{P} = 0$).

Hence the Δ^{-} particle seems to consist of three *down* quarks only: *ddd*

Because the π^{-} meson which is the composition of $\bar{u}d$ quarks is positioned at the third level while the rest of one *up* quark and two *down* quarks of the neutron are positioned at the second and the first levels respectively, there is no conflict with the Pauli Exclusion. The schema below shows it clearly:

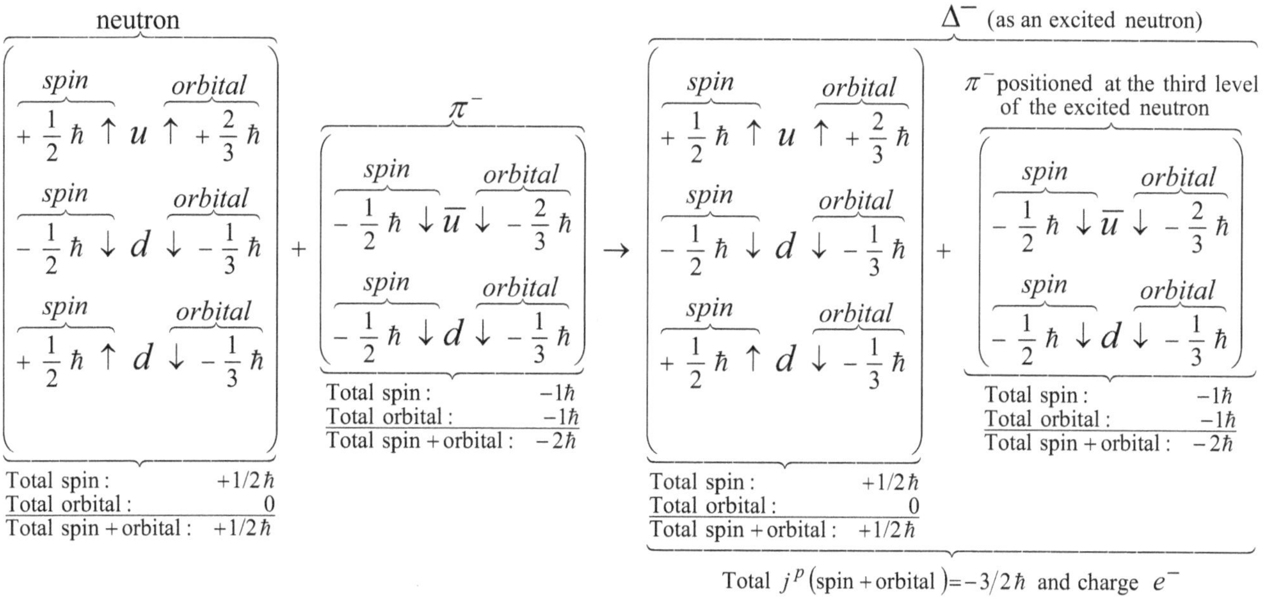

In this mode Δ^{-} decays into the original particles that created it via the strong force: $\Delta^{-} \rightarrow n + \pi^{-}$

* Was the colored quarks classification needed?

The baryons beyond the proton and neutron are basically (before instantly decaying) a composition of a neutron or a proton which consist of a *down* and *up* quarks at their first and second energy levels, and additional quarks that as singles they represent **fermions**, but in the neutron or the proton they form a bound state **bosons** compositions as a charged Pi mesons that are positioned at their third energy level including some neutral Pi mesons too (Please see additional explanation on the actual role of the neutral Pi meson within the proton given for the Omega Minus page 186) hence bypassing the Pauli Exclusion Principle. So <u>from this aspect only</u> the need to invent the colored quarks to comply with the Pauli Exclusion Principle seems needless <u>though gluons do exist</u>!(see page 152). Here are few examples using the new definition of the second and third quarks generations.

1. $$\Sigma^{*\circ}\langle uds\rangle \to ud\underbrace{u\bar{u}d}_{s} \to \underbrace{\underbrace{uud}_{\text{P}}}_{\text{quarks at the levels I \& II of the proton}} + \underbrace{\underbrace{\bar{u}d}_{\pi^-}}_{\text{positioned at level III of the proton}}$$

2. $$\Sigma^{*-}\langle dds\rangle \to dd\underbrace{u\bar{u}d}_{s} \to \underbrace{\underbrace{udd}_{\text{n}}}_{\text{quarks at the levels I \& II of the neutron}} + \underbrace{\underbrace{\bar{u}d}_{\pi^-}}_{\text{positioned at level III of the neutron}}$$

3. $$\Sigma^{*+}\langle uus\rangle \to uu\underbrace{d\bar{d}d}_{s} \to \underbrace{\underbrace{udd}_{\text{n}}}_{\text{quarks at the levels I \& II of the neutron}} + \underbrace{\underbrace{u\bar{d}}_{\pi^+}}_{\text{positioned at level III of the neutron}}$$

4. $$\Omega^{-}\langle sss\rangle \to \underbrace{d\bar{d}d}_{s}\underbrace{u\bar{u}d}_{s}\underbrace{u\bar{u}d}_{s} \to \underbrace{\underbrace{uud}_{\text{p}}}_{\text{quarks at the levels I \& II of the proton}} + \underbrace{\underbrace{\bar{u}d}_{\pi^-} + \underbrace{\bar{u}d}_{\pi^-}}_{\text{positioned at level III of the proton}} + \underbrace{d\bar{d}}_{\pi^\circ}$$

5. $$\Xi_c^{*\circ}\langle dcs\rangle \to d\underbrace{s\bar{s}u}_{c}s \to d\underbrace{d\bar{d}d}_{s}\underbrace{u\bar{u}\bar{d}}_{\bar{s}}u\underbrace{u\bar{u}d}_{s} \to \underbrace{\underbrace{udd}_{\text{n}}}_{\text{quarks at the levels I \& II of the neutron}} + \underbrace{\underbrace{u\bar{d}}_{\pi^+} + \underbrace{u\bar{d}}_{\pi^+} + \underbrace{\bar{u}d}_{\pi^-} + \underbrace{\bar{u}d}_{\pi^-}}_{\text{positioned at level III of the neutron}}$$

6. $$\Omega_c^{*\circ}\langle ssc\rangle \to ss\underbrace{s\bar{s}u}_{c} \to \underbrace{d\bar{d}d}_{s}\underbrace{u\bar{u}d}_{s}\underbrace{d\bar{d}d}_{s}\underbrace{u\bar{u}\bar{d}}_{\bar{s}}u \to \underbrace{\underbrace{udd}_{\text{n}}}_{\text{quarks at the levels I \& II of the neutron}} + \underbrace{\underbrace{u\bar{d}}_{\pi^+} + \underbrace{u\bar{d}}_{\pi^+} + \underbrace{\bar{u}d}_{\pi^-} + \underbrace{\bar{u}d}_{\pi^-}}_{\text{positioned at level III of the neutron}} + \underbrace{d\bar{d}}_{\pi^\circ}$$

7. $$\Xi_b^{-}\langle dsb\rangle \to d\underbrace{u\bar{u}d}_{s}\underbrace{c\bar{c}s}_{b} \to du\bar{u}d\underbrace{s\bar{s}u}_{c}\underbrace{s\bar{s}\bar{u}}_{\bar{c}}s \to du\bar{u}d\underbrace{d\bar{d}d}_{s}\underbrace{u\bar{u}\bar{d}}_{\bar{s}}u\underbrace{d\bar{d}d}_{s}\underbrace{u\bar{u}\bar{d}}_{\bar{s}}\bar{u}\underbrace{u\bar{u}d}_{s}$$
$$\underbrace{\underbrace{udd}_{\text{n}}}_{\text{quarks at the levels I \& II of the neutron}} + \underbrace{\underbrace{u\bar{d}}_{\pi^+} + \underbrace{u\bar{d}}_{\pi^+} + \underbrace{u\bar{d}}_{\pi^+} + \underbrace{u\bar{d}}_{\pi^+} + \underbrace{\bar{u}d}_{\pi^-} + \underbrace{\bar{u}d}_{\pi^-} + \underbrace{\bar{u}d}_{\pi^-} + \underbrace{\bar{u}d}_{\pi^-} + \underbrace{\bar{u}d}_{\pi^-}}_{\text{positioned at level III of the neutron}} \hookleftarrow$$

8. $$\Sigma_b^{*\circ}\langle udb\rangle \to ud\underbrace{c\bar{c}s}_{b} \to ud\underbrace{s\bar{s}u}_{c}\underbrace{s\bar{s}\bar{u}}_{\bar{c}}s \to ud\underbrace{u\bar{u}d}_{s}\underbrace{d\overline{\bar{d}d}}_{\bar{s}}u\underbrace{u\bar{u}d}_{s}\underbrace{d\overline{\bar{d}d}}_{\bar{s}}\bar{u}\underbrace{u\bar{u}d}_{s}$$
$$\underbrace{\underbrace{udd}_{\text{n}}}_{\text{quarks at the levels I \& II of the neutron}} + \underbrace{\underbrace{u\bar{d}}_{\pi^+} + \underbrace{u\bar{d}}_{\pi^+} + \underbrace{u\bar{d}}_{\pi^+} + \underbrace{u\bar{d}}_{\pi^+} + \underbrace{\bar{u}d}_{\pi^-} + \underbrace{\bar{u}d}_{\pi^-} + \underbrace{\bar{u}d}_{\pi^-} + \underbrace{\bar{u}d}_{\pi^-}}_{\text{positioned at level III of the neutron}} \hookleftarrow$$

16

The $W^{\pm}$ Boson of the Weak Force and the True Nature of the Electron and the other Leptons

Previously I have claimed that in the weak decay process, the $\pi^{\pm}$ meson is being emitted out along with a γ photon that carries the released kinetic energy ~80[Gev] from the third level and jointly they form a massive $W^{\pm}$ boson that later decays into leptons. What really occurs is the following:

a. The electron e^{-}:

The π^{-} meson is emitted out along with a γ photon that carries the released kinetic energy ~80[Gev] from the third level of the excited neutron. The π^{-} meson coupled to a γ photon form the W^{-} boson! The W^{-} boson has a charge of e^{-} acquired from the π^{-} meson and a $(-\hbar\downarrow)$ spin acquired from the orbital angular momentum of the π^{-} meson in the excited neutron composition. The W^{-} boson later decays into a π^{-} meson along with an electron neutrino ν_e and electron anti neutrino $\overline{\nu}_e$ that were produced from the γ photon.

The π^{-} meson and the electron neutrino ν_e create a bound state composition $(\pi^{-} + \nu_e)$.

* The bound state composition $(\pi^{-} + \nu_e)$ is **the electron** e^{-} **inner structure (schema page 153)**

The bound state composition $(\pi^{-} + \nu_e)$ that represents the actual electron e^{-} entity along with the electron anti neutrino $\overline{\nu}_e$ are the leptons products of W^{-} boson.

The electron has a charge of e^{-} acquired from π^{-} meson and a total spin of $(-\frac{1}{2}\hbar\downarrow)$ that represents the algebraic sum of the $(-\hbar\downarrow)$ acquired from the orbital angular momentum (OAM) of the π^{-} meson and the $(+\frac{1}{2}\hbar\uparrow)$ spin of the electron neutrino ν_e in the composition!

Note: Electrons in extremely confined space are going through a process named "spin-charge separation". The electron splits in two, a spin carrier particle named 'Spinon' and a charge carrier particle named 'Holon'. In our case, the electron neutrino ν_e acting as a Spinon that carries the spin and the π^{-} meson acting as the Holon that carries the e^{-} charge (originated from its intrinsic OAM).

* **Research published on July 2009 by the University of Cambridge and the University of Birmingham in England, showed that electrons could separate into a Spinon and a Holon.**

The detailed decay process of W^{-} boson is described here:

$$\underbrace{\pi^{-} + \gamma}_{W^{-}} \rightarrow \underbrace{\pi^{-} + \overbrace{\nu_e + \overline{\nu}_e}^{\gamma}}_{W^{-}} \rightarrow \underbrace{\overbrace{\pi^{-} + \nu_e}^{e^{-}} + \overline{\nu}_e}_{W^{-}} \rightarrow e^{-} + \overline{\nu}_e$$

b. The positron e^{+}:

The π^{+} meson is emitted out along with a γ photon that carries the released kinetic energy ~80[Gev] from the third level of the excited proton. The π^{+} meson coupled to a γ photon form the W^{+} boson! The W^{+} boson has a charge of e^{+} acquired from the π^{+} meson and a $(+\hbar\uparrow)$ spin acquired from the orbital angular momentum of the π^{+} meson in the excited proton composition. The W^{+} boson later decays into a π^{+} meson along with an electron neutrino ν_e and electron anti neutrino $\overline{\nu}_e$.

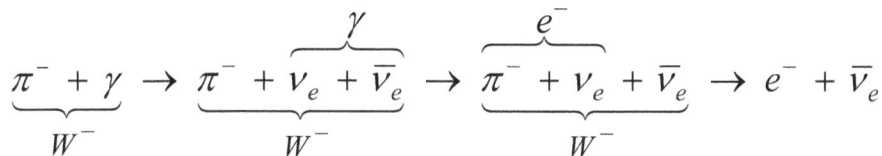

The π^+ meson and the electron anti neutrino $\overline{\nu}_e$ create a bound state composition $(\pi^+ + \overline{\nu}_e)$.

* The bound state composition $(\pi^+ + \overline{\nu}_e)$ is **the positron** e^+ **inner structure**!

The bound state composition $(\pi^+ + \overline{\nu}_e)$ that represents the actual positron e^+ entity along with the electron neutrino ν_e are the leptons products of W^+ boson.

The positron has a charge of e^+ acquired from π^+ meson and a total spin of $(+\frac{1}{2}\hbar \uparrow)$ that represents the algebraic sum of the $(+\hbar \uparrow)$ acquired from the orbital angular momentum of the π^+ meson and the $(-\frac{1}{2}\hbar \downarrow)$ spin of the electron anti neutrino $\overline{\nu}_e$ in the composition!

The detailed decay process of W^+ boson is described here:

$$\underbrace{\pi^+ + \gamma}_{W^+} \rightarrow \underbrace{\pi^+ + \overbrace{\overline{\nu}_e + \nu_e}^{\gamma}}_{W^+} \rightarrow \underbrace{\overbrace{\pi^+ + \overline{\nu}_e}^{e^+} + \nu_e}_{W^+} \rightarrow e^+ + \nu_e$$

c. The **muon** μ^-:

The same process like for the electron is used. The main players are the W^- boson, the π^- meson, the γ photon, the muon neutrino ν_μ and the muon anti neutrino $\overline{\nu}_\mu$, and the muon μ^- itself.

The π^- meson and muon neutrino ν_μ create a bound composition $(\pi^- + \nu_\mu)$.

* The bound state composition $(\pi^- + \nu_\mu)$ is **the muon** μ^- **inner structure**!

The detailed decay process of W^- boson is described here:

$$\underbrace{\pi^- + \gamma}_{W^-} \rightarrow \underbrace{\pi^- + \overbrace{\nu_\mu + \overline{\nu}_\mu}^{\gamma}}_{W^-} \rightarrow \underbrace{\overbrace{\pi^- + \nu_\mu}^{\mu^-} + \overline{\nu}_\mu}_{W^-} \rightarrow \mu^- + \overline{\nu}_\mu$$

d. The **muon** μ^+:

The same process like for the positron is used. The main players are the W^+ boson, the π^+ meson, the γ photon, the muon neutrino ν_μ and the muon anti neutrino $\overline{\nu}_\mu$, and the muon μ^+ itself.

The π^+ meson and muon anti neutrino $\overline{\nu}_\mu$ create a bound state composition $(\pi^+ + \overline{\nu}_\mu)$.

* The bound state composition $(\pi^+ + \overline{\nu}_\mu)$ is **the muon** μ^+ **inner structure**!

The detailed decay process of W^+ boson is described here:

$$\underbrace{\pi^+ + \gamma}_{W^+} \rightarrow \underbrace{\pi^+ + \overbrace{\overline{\nu}_\mu + \nu_\mu}^{\gamma}}_{W^+} \rightarrow \underbrace{\overbrace{\pi^+ + \overline{\nu}_\mu}^{\mu^+} + \nu_\mu}_{W^+} \rightarrow \mu^+ + \nu_\mu$$

e. The **tau** τ^-:

The tau τ^- has four different inner structures! that each of them can have a different meson besides the pi meson that reaches the third level of a relevant baryon, and while these mesons are emitted out from the third level along with a γ photon that carries the released kinetic energy ~80[Gev] they jointly form the W^- boson.

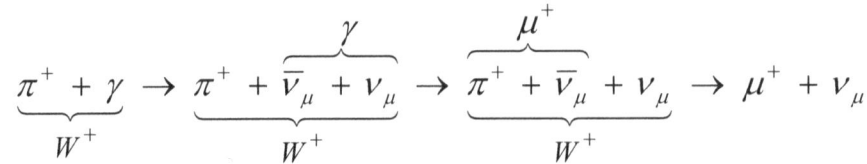

The four mesons are: π^- meson, K^- meson, D_S^- (strange D meson) and B_c^- (charmed B meson).

Note: The component that defines the different tau τ^- lepton inner structure as being identical is the same neutrino that each variant consist of!

1. π^- meson along with a γ photon are emitted out from the third level in a baryon and jointly they form a W^- boson. The W^- boson later decays into a π^-meson along with a tau neutrino ν_τ and a tau anti neutrino $\bar{\nu}_\tau$ that were produced from the γ photon.

The π^- meson and tau neutrino ν_τ create a bound state composition $(\pi^- + \nu_\tau)$.

* The bound state composition of $(\pi^- + \nu_\tau)$ is a **pion-based tau τ^- inner structure**!

The detailed decay process of W^- boson is described here:

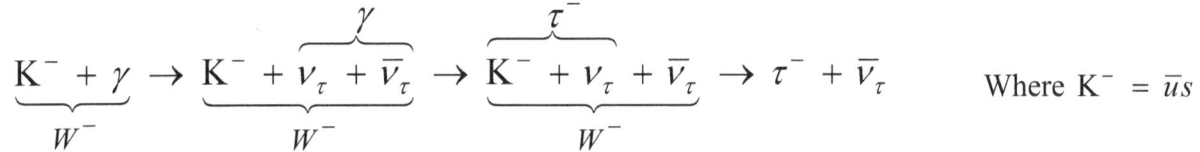

$$\underbrace{\pi^- + \gamma}_{W^-} \rightarrow \underbrace{\pi^- + \overbrace{\nu_\tau + \bar{\nu}_\tau}^{\gamma}}_{W^-} \rightarrow \underbrace{\overbrace{\pi^- + \nu_\tau}^{\tau^-} + \bar{\nu}_\tau}_{W^-} \rightarrow \tau^- + \bar{\nu}_\tau$$

This form of tau τ^- decays into: $\tau^- \rightarrow \pi^- + \nu_\tau$

2. K^- meson along with a γ photon are emitted out from the third level in a baryon and jointly they form a W^- boson. The W^- boson later decays into a K^- meson along with tau neutrino ν_τ and a tau anti neutrino $\bar{\nu}_\tau$ that were produced from the γ photon.

The K^- meson and tau neutrino ν_τ create a bound state composition $(K^- + \nu_\tau)$.

* The bound state composition $(K^- + \nu_\tau)$ is a **kaon-based tau τ^- inner structure**!

The detailed decay process of W^- boson is described here:

$$\underbrace{K^- + \gamma}_{W^-} \rightarrow \underbrace{K^- + \overbrace{\nu_\tau + \bar{\nu}_\tau}^{\gamma}}_{W^-} \rightarrow \underbrace{\overbrace{K^- + \nu_\tau}^{\tau^-} + \bar{\nu}_\tau}_{W^-} \rightarrow \tau^- + \bar{\nu}_\tau \qquad \text{Where } K^- = \bar{u}s$$

This form of tau τ^- decays into: $\tau^- \rightarrow \pi^- + \pi^\circ + \nu_\tau$

Because: $K^- = \bar{u}\underbrace{u\bar{u}d}_{s} \rightarrow \pi^- + \pi^\circ$

3. D_S^- (strange D meson) along with a γ photon are emitted out from the third level in a baryon and jointly they form a W^- boson. The W^- boson later decays into a D_S^- meson along with tau neutrino ν_τ and a tau anti neutrino $\bar{\nu}_\tau$ that were produced from the γ photon.

The D_S^- meson and tau neutrino ν_τ create a bound state composition $(D_S^- + \nu_\tau)$.

* The bound state composition of $(D_S^- + \nu_\tau)$ is a **D meson-based tau τ^- inner structure**!

The detailed decay process of W^- boson is described here:

$$\underbrace{D_S^- + \gamma}_{W^-} \rightarrow \underbrace{D_S^- + \overbrace{\nu_\tau + \bar{\nu}_\tau}^{\gamma}}_{W^-} \rightarrow \underbrace{\overbrace{D_S^- + \nu_\tau}^{\tau^-} + \bar{\nu}_\tau}_{W^-} \rightarrow \tau^- + \bar{\nu}_\tau \qquad \text{Where } D_S^- = s\bar{c}$$

This form of tau τ^- decays into: $\tau^- \rightarrow \pi^- + \pi^- + \pi^+ + \nu_\tau$ or $\tau^- \rightarrow \pi^\circ + \pi^\circ + \pi^- + \nu_\tau$

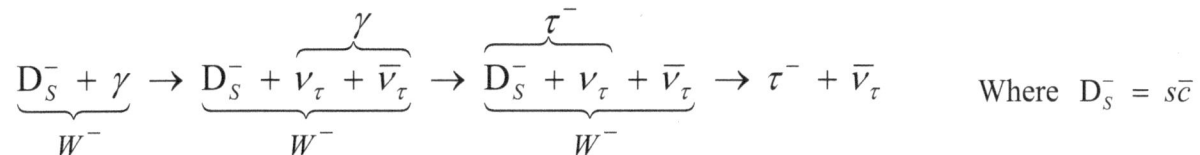

Because: $D_S^- = \underbrace{u\bar{u}d}_{s}\underbrace{d\bar{u}\bar{d}}_{\bar{c}} \rightarrow \pi^- + \pi^- + \pi^+$ or $D_S^- = \underbrace{u\bar{u}d}_{s}\underbrace{d\bar{u}\bar{d}}_{\bar{c}} \rightarrow \pi^\circ + \pi^\circ + \pi^-$

4. B_c^- (charmed B meson) along with a γ photon are emitted out from the third level in a baryon and jointly they form a W^- boson. The W^- boson later decays into a B_c^- meson along with tau neutrino ν_τ and a tau anti neutrino $\bar{\nu}_\tau$ that were produced from the γ photon.

The B_c^- meson and tau neutrino ν_τ create a bound state composition $(B_c^- + \nu_\tau)$.

* The bound state composition of $(B_c^- + \nu_\tau)$ is a **B meson-based tau τ^- inner structure**!

The detailed decay process of W^- boson is described here:

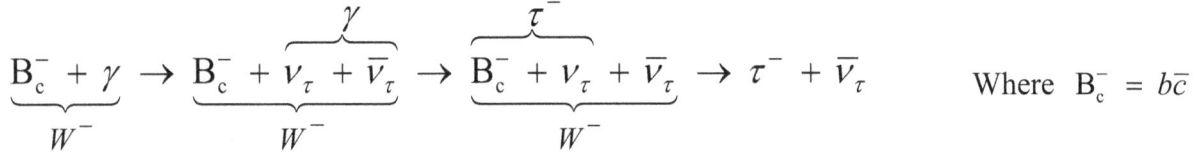

$$\underbrace{B_c^- + \gamma}_{W^-} \rightarrow \underbrace{B_c^- + \overbrace{\nu_\tau + \bar{\nu}_\tau}^{\gamma}}_{W^-} \rightarrow \underbrace{\overbrace{B_c^- + \nu_\tau}^{\tau^-} + \bar{\nu}_\tau}_{W^-} \rightarrow \tau^- + \bar{\nu}_\tau \qquad \text{Where } B_c^- = b\bar{c}$$

This form of tau τ^- decays into:

$\tau^- \rightarrow \pi^\circ + \pi^- + \pi^- + \pi^+ + \nu_\tau$ or $\tau^- \rightarrow \pi^\circ + \pi^\circ + \pi^\circ + \pi^- + \nu_\tau$

Because:

$$B_c^- = \underbrace{u\bar{u}s}_{b}\underbrace{d\bar{d}\bar{u}}_{\bar{c}} \rightarrow \underbrace{u\bar{u}\overbrace{u\bar{u}d}^{s}}_{b}\underbrace{d\bar{d}\bar{u}}_{\bar{c}} \rightarrow \pi^\circ + \pi^- + \pi^- + \pi^+ \text{ or } B_c^- = \underbrace{u\bar{u}\overbrace{u\bar{u}d}^{s}}_{b}\underbrace{d\bar{d}\bar{u}}_{\bar{c}} \rightarrow \pi^\circ + \pi^\circ + \pi^\circ + \pi^-$$

f. The **tau** τ^+:

The tau τ^+ has four different inner structures! that each of them can have a different meson besides the pi meson that reaches the third level of a relevant baryon and while these mesons are emitted out from the third level along with a γ photon that carries the released kinetic energy ~80[Gev] they jointly form the W^+ boson.

The four mesons are: π^+ meson, K^+ meson, D_S^+ (strange D meson) and B_c^+ (charmed B meson).

Note: The component that defines the different tau τ^+ lepton inner structure as being identical is the same neutrino that each variant consist of!

1. π^+ meson along with a γ photon are emitted out from the third level in a baryon and jointly they form a W^+ boson. The W^+ boson later decays into a π^+ meson along with tau neutrino ν_τ and a tau anti neutrino $\bar{\nu}_\tau$ that were produced from the γ photon.

The π^+ meson and tau anti neutrino $\bar{\nu}_\tau$ create a bound state composition $(\pi^+ + \bar{\nu}_\tau)$.

The bound state composition of $(\pi^+ + \bar{\nu}_\tau)$ is a **pion-based tau τ^+ inner structure**!

The detailed decay process of W^+ boson is described here:

$$\underbrace{\pi^+ + \gamma}_{W^+} \rightarrow \underbrace{\pi^+ + \overbrace{\bar{\nu}_\tau + \nu_\tau}^{\gamma}}_{W^+} \rightarrow \underbrace{\overbrace{\pi^+ + \bar{\nu}_\tau}^{\tau^+} + \nu_\tau}_{W^+} \rightarrow \tau^+ + \nu_\tau$$

This form decays into: $\tau^+ \rightarrow \pi^+ + \bar{\nu}_\tau$

2. K^+ meson along with a γ photon are emitted out from the third level in a baryon and jointly they form a W^+ boson. The W^+ boson later decays into a K^+ meson along with tau neutrino ν_τ and a tau anti neutrino $\bar{\nu}_\tau$ that were produced from the γ photon.

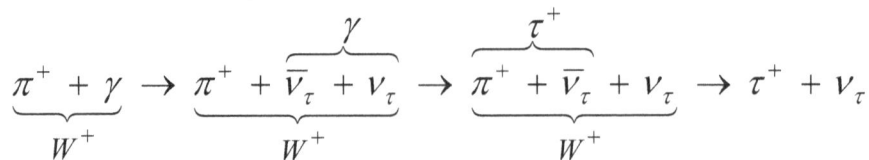

The K^+ meson and tau anti neutrino $\bar{\nu}_\tau$ create a bound state composition $(K^+ + \bar{\nu}_\tau)$.

* The bound state composition of $(K^+ + \bar{\nu}_\tau)$ is a **kaon-based tau τ^+ inner structure**!

The detailed decay process of W^+ boson is described here:

$$\underbrace{K^+ + \gamma}_{W^+} \rightarrow \underbrace{K^+ + \overbrace{\bar{\nu}_\tau + \nu_\tau}^{\gamma}}_{W^+} \rightarrow \underbrace{\overbrace{K^+ + \bar{\nu}_\tau}^{\tau^+} + \nu_\tau}_{W^+} \rightarrow \tau^+ + \nu_\tau \qquad \text{Where } K^+ = u\bar{s}$$

This form of tau τ^+ decays into: $\tau^+ \rightarrow \pi^+ + \pi^\circ + \bar{\nu}_\tau$

Because: $K^+ = u\underbrace{u\bar{u}\bar{d}}_{\bar{s}} \rightarrow \pi^+ + \pi^\circ$

3. D_S^+ (Strange D meson) along with a γ photon are emitted out from the third level in a baryon and jointly they form a W^+ boson. The W^+ boson later decays into a D_S^+ meson along with a tau neutrino ν_τ and a tau anti neutrino $\bar{\nu}_\tau$ that were produced from the γ photon.

The D_S^+ meson and tau anti neutrino $\bar{\nu}_\tau$ create a bound state composition $(D_S^+ + \bar{\nu}_\tau)$.

The bound state composition of $(D_S^+ + \bar{\nu}_\tau)$ is a **D meson-based tau τ^+ inner structure**!

The detailed decay process of W^+ boson is described here:

$$\underbrace{D_S^+ + \gamma}_{W^+} \rightarrow \underbrace{D_S^+ + \overbrace{\bar{\nu}_\tau + \nu_\tau}^{\gamma}}_{W^+} \rightarrow \underbrace{\overbrace{D_S^+ + \bar{\nu}_\tau}^{\tau^+} + \nu_\tau}_{W^+} \rightarrow \tau^+ + \nu_\tau \qquad \text{Where } D_S^+ = \bar{s}c$$

This form of tau τ^+ decays into: $\tau^+ \rightarrow \pi^+ + \pi^+ + \pi^- + \bar{\nu}_\tau$ or $\tau^+ \rightarrow \pi^\circ + \pi^\circ + \pi^+ + \bar{\nu}_\tau$

Because: $D_S^+ = \underbrace{u\bar{u}\bar{d}}_{\bar{s}}\underbrace{\bar{d}ud}_{c} \rightarrow \pi^+ + \pi^+ + \pi^-$ or $D_S^+ = \underbrace{u\bar{u}\bar{d}}_{\bar{s}}\underbrace{\bar{d}ud}_{c} \rightarrow \pi^\circ + \pi^\circ + \pi^+$

4. B_c^+ (Charmed B meson) along with the γ photon are emitted out from the third level in a baryon and jointly they form a W^+ boson. The W^+ boson later decays into a B_c^+ meson along with a tau neutrino ν_τ and a tau anti neutrino $\bar{\nu}_\tau$ that were produced from the γ photon.

The B_c^+ meson and tau anti neutrino $\bar{\nu}_\tau$ create a bound state composition $(B_c^+ + \bar{\nu}_\tau)$.

* The bound state composition of $(B_c^+ + \bar{\nu}_\tau)$ is a **B meson-based tau τ^+ inner structure**!

The detailed decay process of W^+ boson is described here:

$$\underbrace{B_c^+ + \gamma}_{W^+} \rightarrow \underbrace{B_c^+ + \overbrace{\bar{\nu}_\tau + \nu_\tau}^{\gamma}}_{W^+} \rightarrow \underbrace{\overbrace{B_c^+ + \bar{\nu}_\tau}^{\tau^+} + \nu_\tau}_{W^+} \rightarrow \tau^+ + \nu_\tau \qquad \text{Where } B_c^+ = \bar{b}c$$

This form of tau τ^+ decays into:

$\tau^+ \rightarrow \pi^\circ + \pi^- + \pi^+ + \pi^+ + \bar{\nu}_\tau$ or $\tau^+ \rightarrow \pi^\circ + \pi^\circ + \pi^\circ + \pi^+ + \bar{\nu}_\tau$

Because:

$$B_c^+ = \underbrace{u\bar{u}\bar{s}}_{\bar{b}}\underbrace{d\bar{d}u}_{c} \rightarrow \underbrace{u\bar{u}\overbrace{u\bar{u}\bar{d}}^{\bar{s}}}_{\bar{b}}\underbrace{u\bar{d}d}_{c} \rightarrow \pi^\circ + \pi^- + \pi^+ + \pi^+ \text{ or } B_c^+ = \underbrace{u\bar{u}\overbrace{u\bar{u}\bar{d}}^{\bar{s}}}_{\bar{b}}\underbrace{u\bar{d}d}_{c} \rightarrow \pi^\circ + \pi^\circ + \pi^\circ + \pi^+$$

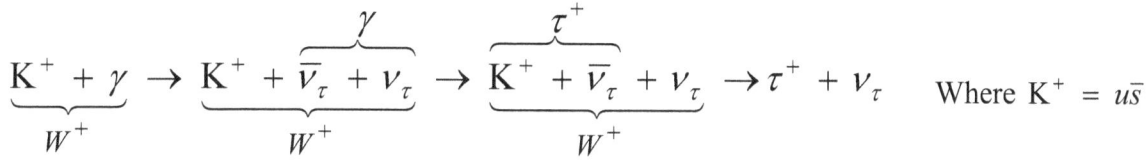

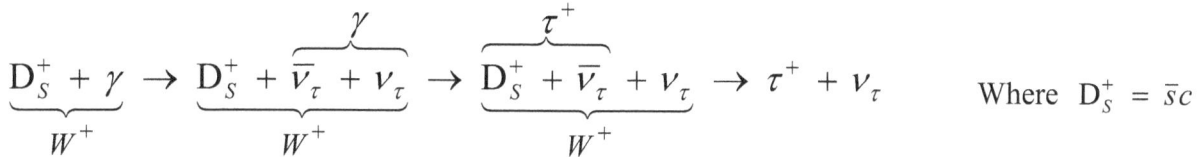

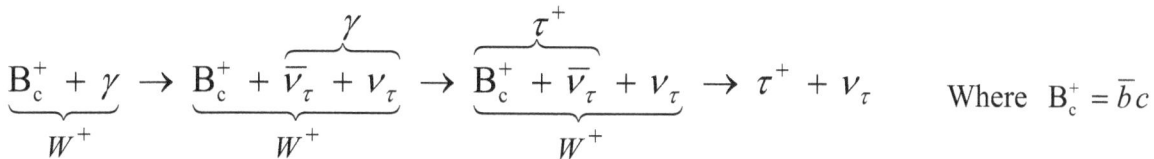

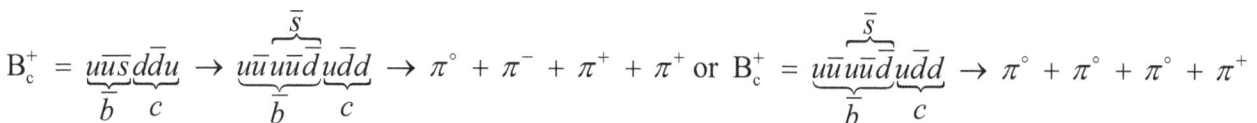

g. The electron collides with a positron:

While an electron collides with a positron the energy of the collision is embodied in a γ photon that provides a source for two neutrinos that are being generated from it, an electron neutrino ν_e and an electron anti neutrino $\bar{\nu}_e$. The electron and the positron that originally collided along with the electron neutrino ν_e and electron anti neutrino $\bar{\nu}_e$ generated from the γ photon yields the following:

$$e^- + e^+ + \overbrace{\nu_e + \bar{\nu}_e}^{\gamma} \rightarrow \underbrace{e^- + \bar{\nu}_e}_{W^-} + \underbrace{e^+ + \nu_e}_{W^+} \rightarrow W^- + W^+$$

h. The Oscillation of neutrinos in electron/ positron collision and in a muon decaying mode:

The neutrinos in the electron and the positron bound states are being switched by the muon neutrinos generated from a γ photon produced from their collision, which form a new bound state in new formed particles as a muon and anti muon (see schema below)!

$$\overbrace{\pi^- + \nu_e}^{e^-} + \overbrace{\pi^+ + \bar{\nu}_e}^{e^+} + \overbrace{\nu_\mu + \bar{\nu}_\mu}^{\text{Generated from a } \gamma \text{ photon}} \rightarrow \overbrace{\pi^- + \nu_\mu}^{\mu^-} + \overbrace{\pi^+ + \bar{\nu}_\mu}^{\mu^+} + \overbrace{\nu_e + \bar{\nu}_e}^{\text{Annihilated into a } \gamma \text{ photon}} \rightarrow \mu^- + \mu^+ + \gamma$$

Another case is of a muon neutrino that initially is in a bound state with a negative Pi meson which forms a muon (produced in the upper atmosphere by cosmic rays), that is being switched by an electron neutrino generated from a γ photon produced in a solar outburst, which forms a new bound state as an electron particle accompanied by the free released muon neutrino and an anti electron neutrino scattered alongside. To an outsider observer it seems that the electron neutrino oscillates between its state and the muon neutrino state causing a deficit in the amount of electron neutrinos that reaches the surface of the earth!

$$\overbrace{\pi^- + \nu_\mu}^{\mu^-} + \overbrace{\nu_e + \bar{\nu}_e}^{\text{a } \gamma \text{ photon from the solar outburst}} \rightarrow \overbrace{\pi^- + \nu_e}^{e^-} + \bar{\nu}_e + \nu_\mu$$

i. The true nature of the Z° boson.

The π^- meson is emitted out along with a γ photon that carries the released kinetic energy ~80[Gev] from the third level of an excited neutron. The π^- meson coupled to a γ photon form the W^- boson which later decays into a π^- meson bound to an electron neutrino ν_e creating the electron particle along with a free electron anti neutrino $\bar{\nu}_e$ that was generated from the γ photon as following:

$$\underbrace{\pi^- + \gamma}_{W^-} \rightarrow \underbrace{\pi^- + \overbrace{\nu_e + \bar{\nu}_e}^{\gamma}}_{W^-} \rightarrow \underbrace{\overbrace{\pi^- + \nu_e}^{e^-} + \bar{\nu}_e}_{W^-} \rightarrow e^- + \bar{\nu}_e$$

And when a π^+ meson is emitted out along with a γ photon that carries the released kinetic energy ~80[Gev] from the third level of an excited proton, the π^+ meson coupled to a γ photon form the W^+ boson which later decays into a π^+ meson bound to an electron anti neutrino $\bar{\nu}_e$ creating a positron particle and a free electron neutrino ν_e that was generated from the γ photon as following:

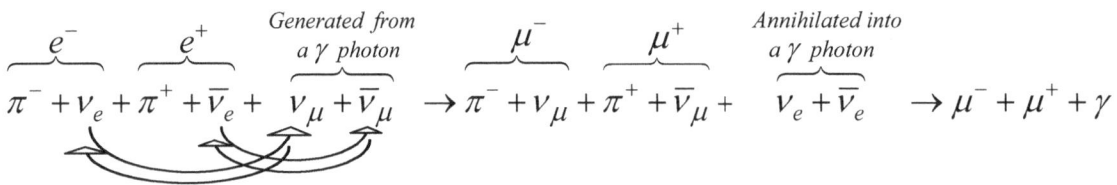

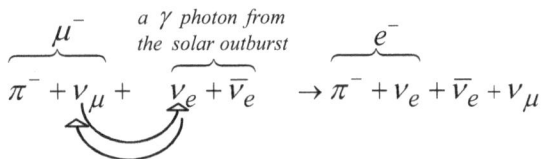

$$\underbrace{\pi^+ + \gamma}_{W^+} \rightarrow \underbrace{\pi^+ + \overbrace{\bar{\nu}_e + \nu_e}^{\gamma}}_{W^+} \rightarrow \underbrace{\overbrace{\pi^+ + \bar{\nu}_e}^{e^+} + \nu_e}_{W^+} \rightarrow e^+ + \nu_e$$

The creation of the electron and positron are through π^- or π^+ mesons that each creates a bound state with the neutrinos ν_e and $\bar{\nu}_e$ respectively, leaving the other $\bar{\nu}_e$ and ν_e neutrinos respectively that were generated from the γ photon scattered alongside separately. differently, the Z° boson is created from a π^- and π^+ mesons that initially are coupled to form a bound state at the third energy level, and later this couple is emitted out along with a γ photon that carries the kinetic energy ~80[Gev] of the third level in decaying process. Each meson in the bound state creates a bound state with the neutrinos ν_e and $\bar{\nu}_e$ respectively, leaving no free neutrinos scattered besides the created electron and positron!

$$\underbrace{(\pi^- + \pi^+)}_{\text{bound state}} + \gamma \rightarrow \underbrace{(\pi^- + \pi^+)}_{\text{bound state}} + \overbrace{\nu_e + \bar{\nu}_e}^{\gamma} \rightarrow \underbrace{\overbrace{\pi^- + \nu_e}^{e^-} + \overbrace{\pi^+ + \bar{\nu}_e}^{e^+}}_{Z^\circ} \rightarrow e^- + e^+$$

Note: I've showed the process involving the electron and the positron cases. Another option is with a γ Photon that carries the kinetic energy ~80[Gev] of the third level that later decays into ν_μ and $\bar{\nu}_\mu$ neutrinos and together with the Pi mesons creates μ and $\bar{\mu}$ leptons.

* Here are four experimental cases that were presented previously which correspond to the subject of the excited proton $P_{excited}$ and excited neutron $n_{excited}$, using the inner structure of the leptons!

a) The Cowan - Reines neutrino experiment (the Inverse beta decay): $\bar{\nu}_e + p \rightarrow n + e^+$;

What actually occurs is the following:

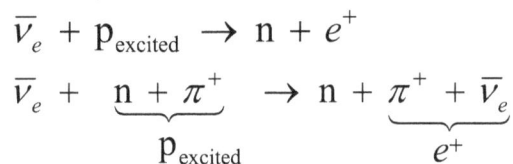

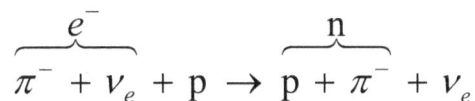

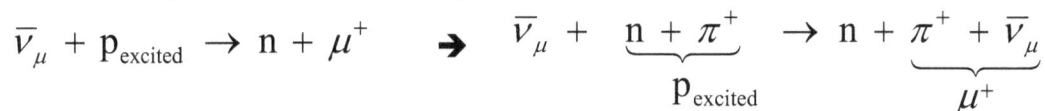

$$\bar{\nu}_e + p_{excited} \rightarrow n + e^+$$

$$\bar{\nu}_e + \underbrace{n + \pi^+}_{p_{excited}} \rightarrow n + \underbrace{\pi^+ + \bar{\nu}_e}_{e^+}$$

b) Electron capture: $e^- + p \rightarrow n + \nu_e$;

What actually occurs is the following:

$$\overbrace{\pi^- + \nu_e}^{e^-} + p \rightarrow \overbrace{p + \pi^-}^{n} + \nu_e$$

c) A muon anti neutrino $\bar{\nu}_\mu$ collides with a proton: $\bar{\nu}_\mu + p \rightarrow n + \mu^+$;

What actually occurs is the following:

$$\bar{\nu}_\mu + p_{excited} \rightarrow n + \mu^+ \quad \Rightarrow \quad \bar{\nu}_\mu + \underbrace{n + \pi^+}_{p_{excited}} \rightarrow n + \underbrace{\pi^+ + \bar{\nu}_\mu}_{\mu^+}$$

d) A muon neutrino ν_μ collides with a neutron: $\nu_\mu + n \rightarrow p + \mu^-$;

What actually occurs is the following:

$$\nu_\mu + n_{excited} \rightarrow p + \mu^- \quad \Rightarrow \quad \nu_\mu + \underbrace{p + \pi^-}_{n_{excited}} \rightarrow p + \underbrace{\pi^- + \nu_\mu}_{\mu^-}$$

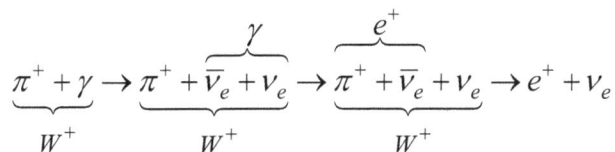

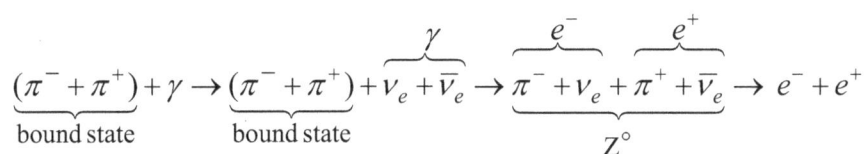

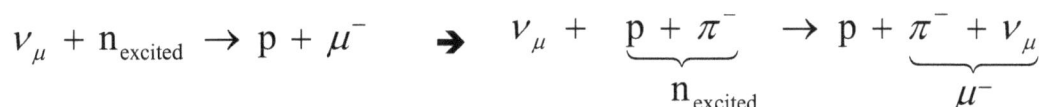

* The π° decaying process that yields 4 leptons, electrons/positrons or muons/anti-muons like it was revealed at the Higgs boson discovery!

Note: It was stated at page 178 that the baryons beyond the proton and neutron are basically (before instantly decaying) a composition of a neutron or a proton with a *down* and *up* quarks at their I&II energy levels, with additional charged Pi mesons that are positioned at their third energy level that may include <u>some neutral Pi mesons too</u>! Following here is an example which uses the new definition of the II&III quarks generations that describes the actual decaying process of the neutral Pi meson following the Omega Minus decay sequence presented at pages 165!

The following schema presents a proton (uud) that its two *up* and one *down* quarks are positioned at the I&II energy levels, and in addition it contains two π^- mesons at its third energy level, and a π° meson that decays in a process described here below. **The golden rule decay mode** of π° resembles the decaying of a couple of π^+ and π^- situated in a 'bound state' positioned at the third energy level:

$$\Omega^-\langle sss\rangle \Rightarrow \underbrace{\bar u d}_{\pi^-}+\underbrace{\bar u u}_{\pi^\circ}+\underbrace{uud}_{\mathrm{P}}+\underbrace{\bar u d}_{\pi^-}\rightarrow \underbrace{\bar u d u}_{s}\bar u u\underbrace{u d\bar u}_{s}d \rightarrow \underbrace{su\bar u}_{b}sd \rightarrow bsd \rightarrow \underbrace{c\bar c d}_{b}sd \rightarrow \underbrace{\underbrace{su\bar d}_{c}\underbrace{\bar s\bar u d}_{\bar c}dsd}_{\downarrow}$$

$$\Omega^-\langle sss\rangle \Rightarrow \underbrace{\bar u d}_{\pi^-}+\underbrace{\overbrace{\underbrace{\bar u d}_{\pi^-}+\underbrace{u\bar d}_{\pi^+}}^{'bound\ state'}+\overbrace{\underbrace{\bar u d}_{\pi^-}+\underbrace{u\bar d}_{\pi^+}}^{'bound\ state'}}_{resembles\ the\ \pi^\circ\ decay\ modes}+\underbrace{uud}_{\mathrm{P}}+\underbrace{\bar u d}_{\pi^-} \quad \leftarrow \quad \underbrace{u\bar u d}_{s}u\bar d\underbrace{u\bar u\bar d}_{\bar s}\bar u dd\underbrace{u\bar u d}_{s}d$$

Note: The added Pi mesons in the process above are from a fused gluons (see page 148 on gluons).

Now, each of the charged mesons couples in the bound state that resembles the π° decaying mode, are being emitted out from the third energy level of the proton accompanied by a γ photon that acquires the energy level and decays into a pair of electron neutrinos $\underbrace{\nu_e+\bar\nu_e}_{\gamma}$ which are bound to a charged Pi meson to yield 4 leptons thru Z° boson (please see Z° formulation at previous page):

$$\underbrace{\underbrace{(\pi^- + \nu_e)}_{e^-}\overset{\swarrow\,\gamma\,\searrow}{+}\underbrace{(\pi^+ + \bar\nu_e)}_{e^+}}_{Z^\circ_{boson}}+\underbrace{\underbrace{(\pi^- + \nu_e)}_{e^-}\overset{\swarrow\,\gamma\,\searrow}{+}\underbrace{(\pi^+ + \bar\nu_e)}_{e^+}}_{Z^\circ_{boson}}$$

Another option of the couples that resembles the π° decaying mode is by being emitted out from the third level of the proton accompanied by a γ photon that acquires the energy level and decays into a pair of muon neutrinos $\underbrace{\nu_\mu+\bar\nu_\mu}_{\gamma}$ that each are bound to a charged Pi meson to yield 4 leptons:

$$\underbrace{\underbrace{(\pi^- + \nu_\mu)}_{\mu^-}\overset{\swarrow\,\gamma\,\searrow}{+}\underbrace{(\pi^+ + \bar\nu_\mu)}_{\mu^+}}_{Z^\circ_{boson}}+\underbrace{\underbrace{(\pi^- + \nu_\mu)}_{\mu^-}\overset{\swarrow\,\gamma\,\searrow}{+}\underbrace{(\pi^+ + \bar\nu_\mu)}_{\mu^+}}_{Z^\circ_{boson}}$$

* Some Examples of Baryons Decay modes

Here are <u>few examples</u> for Xi and Omega Baryons decay modes. The first one Ξ_{cc}^{++} (The Xi Double charmed plus plus particle was discovered on 2017 at CERN!).
The process uses the Gluons fusion that produces the Strange/Anti Strange $s\bar{s}$ (φ) or the Charm/Anti Charm $c\bar{c}$ (J/Ψ) particles during the collision of protons in the particles accelerators that in turn they assist the decay modes of the new particles created in them that has a very short life time (Please refer to the explanation given at page 152).
This process can be used to obtain any particle's decaying modes, sometimes more than a one option for some if continuing the resolution to obtain additional different modes! (Please notice that the new interpretation of the second and third quarks generations is used here!)

$$\Xi_{cc}^{++}(ucc) \rightarrow ucc+s\bar{s} \rightarrow uc\underbrace{u\bar{u}u}_{c}\,s\underbrace{\bar{d}d\bar{d}}_{\bar{s}} \rightarrow \underbrace{udc}_{\Lambda_c^+}+\underbrace{\bar{u}s}_{K^-}+\underbrace{u\bar{d}}_{\pi^+}+\underbrace{u\bar{d}}_{\pi^+}$$

$$\Xi_{cc}^{+}(dcc) \rightarrow dc+\underbrace{s\bar{s}u}_{c} \rightarrow dcs\underbrace{\bar{u}u\bar{d}}_{\bar{s}}\,u \rightarrow \underbrace{udc}_{\Lambda_c^+}+\underbrace{\bar{u}s}_{K^-}+\underbrace{u\bar{d}}_{\pi^+}$$

$$\text{or} \rightarrow \Xi_{cc}^{+}(dcc) \rightarrow dc+\underbrace{s\bar{s}u}_{c} \rightarrow dcs\underbrace{\bar{u}u\bar{d}}_{\bar{s}}\,u \rightarrow \underbrace{uud}_{p}+\underbrace{c\bar{d}}_{D^+}+\underbrace{\bar{u}s}_{K^-}$$

$$\Xi_{b}^{-}(dsb) \rightarrow ds+\underbrace{c\bar{c}d}_{b} \rightarrow \underbrace{dds}_{\Xi^-}+\underbrace{c\bar{c}}_{J/\Psi}$$

$$\Xi_{b}^{0}(usb) \rightarrow u\underbrace{d\bar{d}d}_{s}\,b \rightarrow \underbrace{udb}_{\Lambda_b^0}+\underbrace{d\bar{d}}_{\pi^0}$$

$$\Omega_{b}^{-}(ssb) \rightarrow ss\underbrace{c\bar{c}s}_{b} \rightarrow \underbrace{sss}_{\Omega^-}+\underbrace{c\bar{c}}_{J/\Psi}$$

$$\Omega^{-}(sss) \rightarrow ss\underbrace{u\bar{u}d}_{s} \rightarrow \underbrace{uds}_{\Lambda^0}+\underbrace{\bar{u}s}_{K^-}$$

$$\text{or} \rightarrow \Omega^{-}(sss) \rightarrow ss\underbrace{u\bar{u}d}_{s} \rightarrow \underbrace{uss}_{\Xi^0}+\underbrace{\bar{u}d}_{\pi^-}$$

$$\text{or} \rightarrow \Omega^{-}(sss) \rightarrow ss\underbrace{u\bar{u}d}_{s} \rightarrow \underbrace{dss}_{\Xi^-}+\underbrace{\bar{u}u}_{\pi^0}$$

Summary

The theory presented in this book gives way to unconventional sources from which the fundamental constants of physics may rise. The formulations of the final expressions producing the values that match the experimental ones are logically structured and demonstrate a mathematical symmetry.

My theory generates results that accurately agree with the experimental measurements, based on mathematical foundations. My theory also predicts the possibility of a greater speed of light in the atomic domain, barring consequences on the special theory of relativity and the idea of quantum uncertainty. Provided that the theory is valid, it also unites various areas of physics, from the quantum level to the cosmological level, while clarifying their shared sources. All are indicators of a good theory that should be considered.

I mentioned the experiment based on the EPR paradox[7] when I presented the feasibility for the higher speed of light in the atomic domain. This experiment is covered in detail in Quantum Physics by Issaschar Unna[9]. the experiment mentioned in this book, showed that some elements, like calcium, emit a pair of photons at the same time with identical polarity. one of them travels to the right, and the other one to the left. The experiment also showed that a change made to the polarity direction of the right photon caused a similar change in the left photon, even when the photons were 13 meters apart or more. This means that the left photon somehow knew of the change occurring to the right photon. To justify the possible greater speed of light in the atomic domain, an experiment could be performed that would measure the time interval between the moment of change in the right photon and the moment of change in the left photon. This would allow for an establishment of the speed of information transfer, based on the known distance (13 meters or more). My guess is that the speed will be c_n, the speed of light in the atomic domain, proving beyond any doubt that a greater speed than the speed of light in a vacuum does exist, and

This theory reveals non local Hidden Variables which are not excluded by the Bell's inequality theorem!

In this new revision there are new topics added like the Quarks, the $W^{\pm}$ boson of the weak force decay explained by the three energy levels within the baryons, and a new interpretation of the second and third generation quarks (strange, charm, bottom and top) involving interaction between the subatomic particles. This new revision also presents a new angle on the true nature of the electron and the other leptons by revealing their inner structure!

The physics world is a fascinating place to visit, and I have become, over the years, a persistent traveler. I hope that my theory, if accepted, will serve as a foundation for yet more developments and exciting exploration of this world.

Nomenclature

A	Mass number
A	Cross-sectional area (m^2)
a_e	Bohr radius (m)
$\tilde{a}_e$	Initial Bohr radius (m)
c_n	Speed of light in the atomic domain (ms^{-1})
c	Speed of light in vacuum (ms^{-1})
E_k	Kinetic energy (J)
E_p	Potential energy (J)
e	Elementary charge (C)
$\tilde{e}$	Initial Elementary charge (C)
F	Faraday constant (C $mole^{-1}$)
G	Gravitation constant ($m^3\ s^{-2}\ Kg^{-1}$)
$\tilde{G}$	Initial gravitation constant ($m^3\ s^{-2}\ Kg^{-1}$)
h	Planck constant (J s)
$\tilde{h}$	Initial Planck constant (J s)
h_n	Planck constant in the atomic domain (J s)
$\hbar$	Planck constant divided by 2π (J s)
K	Kepler constant ($s^2\ m^{-3}$)
$\tilde{K}$	Initial Kepler constant ($s^2\ m^{-3}$)
k	Boltzmann constant (J/K^o)
L	Length of contraction (m)
L_o	Rest Length (m)
l_{pl}	Planck length (m)
$\tilde{l}_{pl}$	Initial Planck length (m)
M_s	Sun mass (Kg)
$\tilde{M}_s$	Initial sun mass (Kg)
m_e	Electron mass (Kg)
$\tilde{m}_e$	Initial electron mass (Kg)
m_n	Neutron mass (Kg)
m_p	Proton mass (Kg)

$\widetilde{m}_p$	Initial proton mass (Kg)
$\widetilde{m}_{pl}$	Initial Planck mass (Kg)
$\vec{M}$	relevant magnetic moment of the particle
m_u	Mass number
N_A	Avogadro constant (gmoles^{-1})
P	Pressure (atm)
R	Molar gas constant (J/K^{o} mole)
R_o	Contraction radius (m)
R_H	Hall resistance (Ω)
R_∞	Rydberg constant (m^{-1})
r_n	Neutron radius (m)
r_n'	Relativistic neutron radius (m)
r_p	Proton radius (m)
r_p'	Relativistic proton radius (m)
$\widetilde{r}_p$	Initial proton radius (m)
T_o	Absolute temperature (K^{o})
t_{pl}	Planck time (s)
V	Volume (m^3)
v_e	Electron velocity (ms^{-1})
$\widetilde{v}_e$	Initial electron velocity (ms^{-1})
v_p	Proton spinning velocity (ms^{-1})
r_{n}	Inner Proton's or Neutron's orbital level depends on the orbital number **n**
$\vec{L}$	quantum angular momentum

subscripts

p	Proton
e	Electron
n	Neutron
pl	Planck
C	Compton
H	Hall
A	Avogadro

α	Fine structure constant
$\tilde{\alpha}$	Initial fine structure constant
β	The key constant
γ	Lorentz transformation Factor
δ	Electron magnetic moment
$\varepsilon_\circ$	Permittivity of vacuum constant ($F\ m^{-1}$)
$\tilde{\varepsilon}_\circ$	Initial permittivity of vacuum constant ($F\ m^{-1}$)
η	Ratio between the initial permittivity of vacuum constant and its known value, and between the initial permeability of vacuum constant and its known value
$\lambda_{C,e}$	Compton wavelength of the electron (m)
$\lambda_{C,n}$	Compton wavelength of the neutron (m)
$\lambda_{C,p}$	Compton wavelength of the proton (m)
$\mu_\circ$	Permeability of vacuum constant (Hm^{-1})
$\tilde{\mu}_\circ$	Initial Permeability of vacuum constant (Hm^{-1})
π	Natural number pi
ψ	Proton magnetic moment
φ	Neutron magnetic moment
q_{quark}	Quark's electrical charge
$W^{\pm}$	Weak force boson

Reference

1. Pagels, H.R. *The Cosmic Code: Quantum Physics as the Language of Nature*. New York: Bantam Books, 1983.

2. Halliday, D., and Resnick, R. *Physics, Parts I and II Combined,* Third Edition. New York: John Wiley & Sons, 1978.

3. Kaye, G.W.C., and Laby, T.H. *Tables of Physical and Chemical Constants and Some Mathematical Functions*, Fourteenth Edition. London: Longman Group Limited, 1973.

4. Beiser, A. *Modern Physics: An Introductory Survey*. Cambridge, Massachusets: Addison-Wesly Publishing Company, 1973.

5. Hawking, S. *The Universe in a Nutshell*. New York: Bantam Books, 2001.

6. Feynman, R.P. *QED: The Strange Theory of Light and Matter*. Princeton, New Jersey: Princeton University Press, 1988.

7. Rohrlich, F. *From Paradox to Reality: Our Basic Concepts of the Physical World*. Cambridge, Massachusets: Cambridge University Press, 1987.

8. Cohen, E.R. *The Physics Quick Reference Guide*. Woodbury, New York: American Institute of Physics, 1996.

9. Unna, I. *Quantum Physics*. Tel-Aviv, Israel: Ministry of Defence publication, 1993 (Hebrew).

10. Ne'eman Y & Kirsh Y. *The Particles Hunters: The search after the Fundamentals constituents of Matter*; Israel: Masada, 1983.

11. Yoram Kirsh. *The Universe of the Modern Physics*: Am Oved Publisher Ltd. Tel Aviv 2006.

www.ingramcontent.com/pod-product-compliance
Lightning Source LLC
Chambersburg PA
CBHW041154220326
41598CB00045B/7423
* 9 7 8 1 5 3 4 2 9 8 5 9 0 *